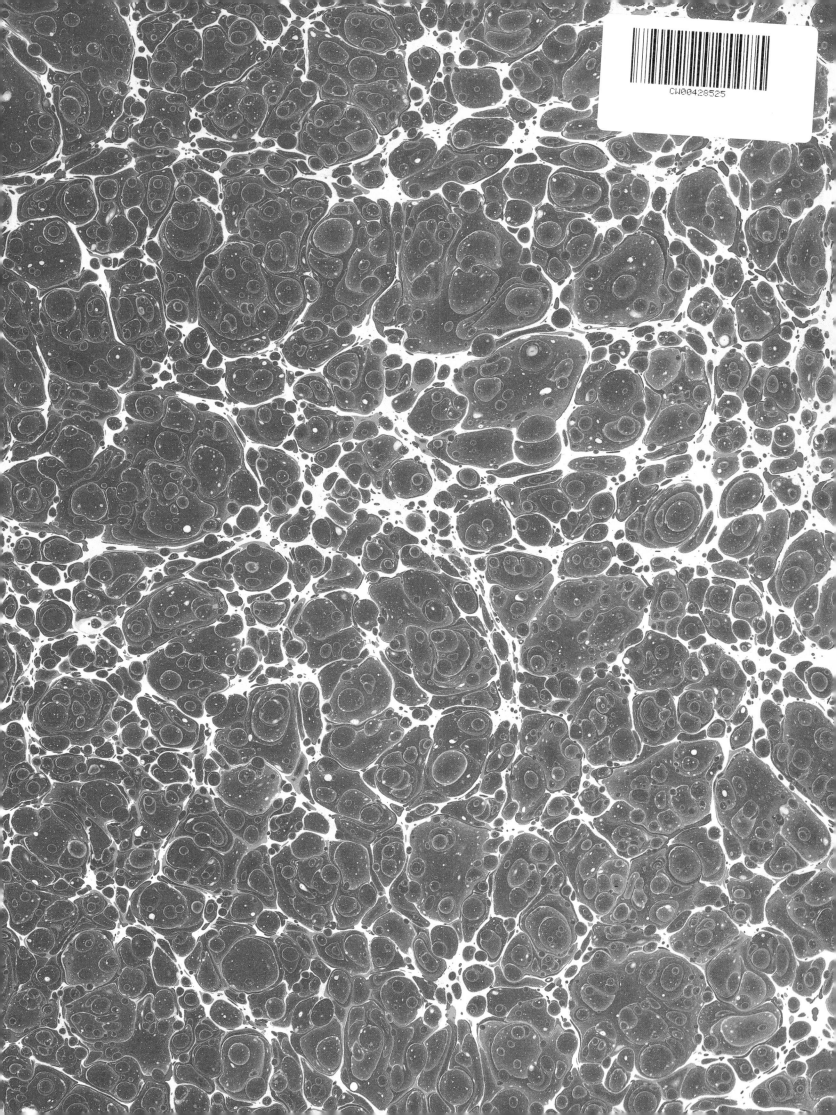

Continental and American
Skeleton Clocks

Continental and American
Skeleton Clocks

Derek Roberts

1469 Morstein Road, West Chester, Pennsylvania 19380

DEDICATION

This book is dedicated to all those craftsmen, both past and present, who have devoted their lives to the restoration and preservation of all the fine clocks produced by their forbears. In particular I would like to express my thanks and appreciation to all the clockmakers, cabinet makers, dial restorers, gliders and the many other people who have assisted me so superbly in this task over last twenty-one years.

Published by Schiffer Publishing, Ltd.
1469 Morstein Road
West Chester, Pennsylania 19380
Please write for a free catalog
This book may be purchased from the publisher.
Please include $2.00 postage.
Try your bookstore first.

Printed in the United States of America.
ISBN: 0-88740-182-1

Acknowledgements

Some seven years ago I was asked by John Steel of The Antique Collector's Club, to produce a book on skeleton clocks. This I readily agreed to do, but by the time it was completed after much research some five years later, it was realized that even after pruning out all superfluous material and reducing it to the minimum compatible with a full coverage of the subject that it would still come out at around 550 pages, which is too large for many people's tastes and pockets. Moreover, someone who is fascinated by British skeleton clocks, for instance, may well have no interest in those produced on the Continent which are markedly different. It was thus decided to divide the book into two: a) *British Skeleton Clocks* which was published by The Antique Collector's Club in October 1987 and b) the current volume covering those produced on the Continent and in America, with a chapter on modern craftsmanship.

It was originally anticipated that The Antique Collector's Club would publish this book but unfortunately production difficulties meant that there would have been along delay before it could reach the bookshops, which would have been undesirable for a variety of reasons. It was at this stage that Peter Schiffer kindly stepped into the breach and offered, if I wished, to get it into print within the original time span, something which I am extremely grateful for. I also greatly appreciate John Steel's generosity in releasing me from my original contract with him and thus enabling me to pass the manuscript over to Schiffer Publishing.

Because of the division of the skeleton clock book into two sections, there is some overlap in the acknowledgements due in the two volumes and thus I would like to reiterate my thanks contained in *British Skeleton Clocks* to the following:

All those people or organizations who kindly either supplied pictures of their clocks or allowed me to photograph them. These included Mr. Albert Odmark, Mr. Norman Langmaid, Dr. S.P. Lehv, Mr. Dudley Heathcote who supplied many photographs of the clocks which were in the collection of his father, Major Anthony Heathcote, Mr. C.H. Bailey, Managing Director of the American Clock and Watch Museum, The "Time Museum" at Rockford, U.S.A., and the major branches of the principal auction houses including Sotheby's, Christie's, and Phillips both in this country and abroad, and various museums.

One must also not forget the contributions of the numerous people who sadly, for reasons of security, can no longer be mentioned here.

I still remember with gratitude the kindness of the staff of the Sammlung Sobek Collection in opening up the Gaymuller Schlossl just outside Vienna for me one cold winter day, and giving me every possible assistance in examining the many fine pieces in their collection, even though the temperature within the building was near freezing. No less helpful was the staff of the Museum at Schoonhoven in Holland when I visited them shortly afterwards.

Professor Hans von Bertele, sadly no longer with us, gave me much help and advice when I visited Vienna and introduced me to several people who have subsequently been of assistance. Following his death, the privilege which the family extended to me in asking me to catalog and advise on the collection was of material help to me in the preparation of the book, particularly so far as Austrian skeleton clocks are concerned.

In France Monsieur Jean Claude Sabrier kindly supplied much useful information and photographs of various skeleton clocks and Madam Catherine Cardinal gave me free access to the superb collection of clocks contained in the Conservatoire National des Arts et Metiers. In Holland I greatly appreciated the late Mr. E. Stender's generosity in allowing me to search through his photographic records and use any of the material I found helpful. Whilst nearer home Mr. Charles Allix provided much useful information and many helpful leads.

I shall always be grateful to the quietly spoken American who met me at Los Angeles Airport some seven years ago, took me back to his house and over the next few days introduced me to practically everyone who possessed a skeleton clock within a hundred mile radius of that city.

Those who gave advice and assistance on specific areas are:

a) Mr. F. Betrand of the Belgian Tourist Office and Mr. Jognery of The School of Horology in that country who, with Mr. Albert Odmark, assisted in researching the work of Hubert Sarton. Mr. Edward G. Aghib's article which appeared in *Antiquarian Horology* was also of great help in this respect.

b) Senor Luis Montanes who supplied details of the skeleton clocks produced by Gutierrez in Spain.

c) Mr. O.R. Hagan formerly proprietor of the Hagans Clock Museum who supplied photographs of various clocks which had been in his collection.

d) Mr. James W. Gibbs who advised on American skeleton clocks and allowed me to quote from his article.

e) The N.A.W.C.C. who assisted me in tracking down various clocks and collectors and advising on a variety of matters.

A considerable number of the clocks illustrated in this book have passed through our hands over the last twenty years and have been photographed down here at Tonbridge and the majority of the remainder have been photographed by the author in the various private collections and museums visited. In this respect I am very grateful to Sky Photographic for all the care they have taken with the numerous films they have processed and printed on my behalf. In addition Narbutt Lieven and Company provided some excellent pictures of the clocks in the Sobek collection for which I am very grateful and Kenneth Clark is responsible for several of the photographs taken in this country including some of those taken at Sotheby's, for whose co-operation I am indebted; and the one wheel clock by Le Roy in The British Museum.

David Penny has kindly made fine line drawings of the escapement on Crane's torsion clock and one of Berthoud's, and John Martin has produced an excellent series of line drawings of remonotoires so as to make this fascinating but complex subject more intelligible.

My thanks as ever are due to my wife and son for their understanding and the support given by them during the production of this book, and I would like to extend this thanks to all my staff who have invariably assisted in any way possible as and when required.

Lastly it would be remiss of me to conclude without thanking all the staff of Schiffer Publishing for the excellent job that they have made of this book's production.

Modern Craftsmanship

At the end of this book we have included a chapter on modern craftsmanship, to show that there are still many fine clockmakers, both professional and amateur, who are producing some beautiful and ingeniously designed and executed pieces.

It is hoped that this will encourage others to enter this field and, at least to some extent, rebuff the cry that we hear all too often "They couldn't make it like that today". "They" can and "they" do, albeit in somewhat limited numbers. However, one suspects that this is more the result of the lack of demand for fine craftsmanship and the unwillingness of people to pay for it in this throw away age, than the potential availability of skilled craftsmen; be they clockmakers, engravers or cabinet makers, to name but a few of the crafts.

Contents

1
Introduction

The term skeleton clock may be taken to include any clock which has been designed with the main purpose of displaying the movement as completely as possible. In practice this usually means making the plates (or frames as they are probably more aptly termed in this context) as delicate relative to the movement as practicable, or fretting them out so that in effect one can see through them. Whereas, with a bracket or longcase clock the movement is enclosed in a wooden case, thus concealing it from view, with a skeleton clock it is covered by a glass dome or sometimes a glazed brass frame and can thus be examined from all angles. Other ways which are sometimes used to assist in the display of the movement are the omission of the dial center, the fretting out of the remaining chapter ring and on occasions the use of glass dials.

Different approaches were used in different countries, for instance in France and Austria a solid inverted Y frame was commonly employed which was very rarely skeletonized, whereas in England the ornamental nature of the frame was a very important consideration when designing the clock and thus the frames were fretted out in many different ways, varying from a single rafter design to fine arabesque scrollwork or even in such a manner as to depict famous buildings. Similar remarks apply to the dial in that on the Continent dials were usually kept solid or with just the center omitted, whereas in England, particularly in the last half of the nineteenth century, the center was nearly always left out and indeed in the majority of cases the remaining chapter ring was also extensively skeletonized.

An interesting technique used in France was the employment of a glass plate for the frame, which many would consider to be the ultimate in skeleton clock design, as it permits one to look through it and see at a glance all the mechanical features of both the front and the back of the clock.

Many of the very early clocks such as the turret and lantern clocks, particularly those made in France, Italy and Germany, were heavily skeletonized purely on the grounds of simplification of construction and economy in the use of materials, but it is doubtful if skeletonizing was ever used as a means of displaying the movement. Some early makers undoubtedly produced clocks which they skeletonized to show off their craftsmanship as completely as possible or laid the clock out in such a way as to draw attention to a fine ingenious mechanical feature. Just such a clock is the lovely little portico clock with cross beat made by Franciscus Schwarz shown on the following page.

Skeletonized Clocks

One type of clock which, while not conceived as a skeleton clock, does deserve a mention here, is a clock of relatively conventional design such as the Viennese regulator, which has been partly skeletonized to display as much of the movement as possible while still maintaining the normal clock case, which enabled the owner to see the details of a complex movement, or an unusual escapement or strike work.

A Gothic iron chamber clock, probably made in the late 16th century, with vertical verge escapement controlled by a balance, internally toothed countwheel and a counter-balanced iron hour hand. Height: 20'' (51 cm).

A small portico clock signed 'Franciscus Schwarz in Bruessel' ca. 1630, which has an engraved bronze and fire gilt frame with some silver ornamentation. The back plate has cut-outs and the clock is laid out in such a way that all the mechanical components can be seen as clearly as possible, with the cross beat escapement, on which the arms have been renewed, mounted prominently at the top. Height: 6 3/4'' (17.2 cm). H. von Bertele Collection.

FRANCE

It is probable that the skeleton clock first appeared in France in the mid-eighteenth century from the gradual evolution of the magnificent pendules de cheminée which were being produced at that time.

Whereas in England much of the finest clockmaking was carried out in the late seventeenth or early eighteenth centuries, the French achieved their best work from 1750-1830 with the arrival on the scene of such brilliant makers as Bertoud, Lepine, the Lepaute family, Janvier and Breguet to name but a few. It was the desire of makers such as these to display their superb craftsmanship to the best possible advantage, which probably gave rise to the skeleton clock, reinforced by the wish of the wealthy few to show off their magnificent objets d'art as impressively as possible. It must be remembered that this was still an age of patronage and undoubtedly many of the finest clockmakers at that time, as in previous centuries, relied heavily on the support of the aristocracy.

In France the number of people able to acquire a fine clock was strictly limited, but those who could afford to buy such a piece demanded only the best and this is why the French skeleton clocks, particularly those made in the eighteenth century, although relatively few in number compared to the English skeleton clocks, were of magnificent quality and frequently of considerable complexity. The French clocks incorporated calendarwork, made use of mechanical refinements such as remontoires and were often fitted with beautifully executed gridiron pendulums. The only skeleton clocks produced in appreciable quantities in France were the very attractive little clocks made in a variety of forms in the mid-nineteenth century at the time of the Great Exhibitions in both London and Paris.

One type of French skeleton clock produced in considerable but far fewer numbers than those just mentioned were the great wheel clocks, mostly timepieces, made at the beginning of the nineteenth century.

AUSTRIA

Skeleton clocks first appeared in Austria, mainly in and around Vienna, towards the end of the eighteenth century, and although they have a superficial resemblance to those made in France, because of the close links between the two countries at that time, they are in effect very different, due in large measure to the somewhat limited financial resources available in Vienna compared with France. Thus, while the clocks were often of great technical merit, they were, with a few notable exceptions, much simpler in conception. Remontoires and true gridiron pendulums were only used occasionally and the clocks were generally much smaller than their French counterparts and had thinner plates, but their relative simplicity and great delicacy of construction give them a charm all of their own. The early Austrian clocks are undoubtedly the most attractive, those produced after the mid-nineteenth century being much heavier in concept and more stereotyped in their design.

ENGLAND

Whereas in France the skeleton clock as an entity gradually evolved over an appreciable period of time, in England the skeleton clock seems to have appeared almost spontaneously c.1820, although a few highly specialized examples such as the one produced by Merlin were constructed in the previous century.

The early English skeleton clocks copied the inverted Y frame, used so extensively in France, but rapidly assumed their own identity and the simple scroll frame became the most popular design. These clocks were probably made in fairly small quantities by an appreciable number of different clockmakers, and it is likely that in the case of these relatively early pieces that the name which appears on the clock is often also that of the maker.

During the period 1820-50, a few fine makers such as Condliff and the partnerships of Strutt and Wigston, and Parker and Pace produced some beautiful quality and highly ingenious clocks, but only in limited numbers. By the mid-nineteenth century the production of skeleton clocks had increased dramatically, roughly matching the rapidly increasing industrial wealth of the country, and it was at this stage that their manufacture was largely taken over by a few specialized firms such as Smith of Clerkenwell and Evans of Handsworth.

Two and three train clocks began to appear in increasing numbers, together with more complex frames (for instance those depicting well-known buildings) and dials became far more decorative. The heyday of skeleton clock production in England was probably in the thirty to forty years following the Great Exhibition of 1851. By 1890 production was slowing down and by 1910 had almost ceased as the populace swung away from what it considered to be the over ornate and excessively elaborate products of the Victorian era, and adopted the classic simplicity of the Edwardian period they had entered, inspired at least in some degree by the beautifully elegant furniture designs produced by Sheraton over a century earlier.

Although the vast majority of skeleton clocks were made in England, France and Austria, some were also produced in limited numbers in other countries such as Belgium, Holland, Spain and America. No trace has been found of skeleton clocks having been manufactured in Italy, which is somewhat surprising in view of its association with the Austro-Hungarian Empire, nor is there information about any originating from Germany. It is likely that a few were commissioned to be made in Switzerland (it is thought that Breguet had some manufactured there) but so far no records exist of any bearing a Swiss clockmaker's name.

2

France

EARLY CLOCKS

The French skeleton clocks, the earliest of which predate their English equivalents by some fifty to sixty years, gradually evolved from the mantel clock and thus the margin between the two is a very blurred one. This evolution probably commenced in the mid 18th century and was stimulated in large measure, by the arrival at that time or shortly after, of a steadily increasing number of French clockmakers of exceptional talent such as the Lepautes, Lepine, the Le Roys, Robin, Janvier, Berthoud, Bouchet, Motel and Breguet. They were all making pieces of great quality and ingenuity and often wished to display these as completely as possible, to enable their customers to appreciate just how fine their work was. One step in this direction was the production of the beautiful series of table regulators usually contained in glazed gilt metal or mahogany cases, so that the movement could readily be seen. These were frequently weight driven or if a spring was chosen as the motive power then a remontoire was often employed so as, in effect, to convert them into a weight driven clock, which is probably the ideal so far as accurate timekeeping is concerned.

Some of these, such as those seen here signed by Lepine and Breguet were fully skeletonized and placed under a glazed cover so that the clockmakers ingenuity was displayed as completely as possible. Many of these were simple timepieces with the sole objective of keeping accurate time, but in other instances various complexities such as Solar and Mean time, sunrise/sunset, lengths of daytime and night-time, lunar calendar and full perpetual calendar were added, and examples of these may be seen in fully skeletonized form throughout this section of the book.

Certain makers such as Verneuil (Figs. 37, 38, 41, 42) who produced an interesting series of skeleton clocks of various degrees of complexity, but all of a similar character, seem to have specialized in this field and usually signed their products, but in other instances, as with the English skeleton clocks, it is often difficult to know who the actual maker of a particular clock is. For instance, the clocks signed by Breguet, Lepine and possibly some of the others in this group, are so similar in concept and construction that one feels that they were probably all made by the same person. It is known that Breguet commissioned many pieces from other makers and it may be that Lepine supplied the clocks under discussion to Breguet. Likewise, it is thought that the three wheel clocks (Fig. 53) of Breguet may well have been made by one of his colleagues in Switzerland.[1]

Early in the 19th century, various series of skeleton clocks were produced in France which, while not made in large numbers and not necessarily exactly the same in every instance, all appear to have come from the same or closely associated workshops. Examples of this are the fascinating glass plated skeleton clocks (Figs. 66-70) with their exquisitely delicate wheelwork, all of which bears a close family resemblance. Other examples (text continued on page 18)

Fig. 1. A clock which illustrates, to some degree, the transition between mantel and skeleton clocks. It still has many of the best decorative features of French 18th century mantel clocks, but has a skeletonized movement and fine wheelwork which can easily be seen beneath the protective glass cover, not shown here.

The movement and dial of this clock in effect form the bob of the gridiron pendulum which has knife edge suspension. The main dial has concentric minute, hour and day hands with the ruling planet being shown, and above this is the thirty-one day fly back calendar which bears the signature Waltrin à Paris and is dated 1781. The miniature in goache is of a classical sacrificial scene in the style of de Gault. Height: 19¾'' (49 cms.) Photo courtesy Christies, London.

---

[1] Daniels G. *The Art of Breguet*, page 80, Sotheby Parke Berner, London & New York, 1975.

Figs. 2a, b A French great wheel skeleton clock decorated with mounts based on a military theme. Photos courtesy Stender B.V. Holland.

Fig. 3. A small and fine quality fully skeletonised Pendule a Cercles tournants (Clock with revolving dials) with enamelled hour and minute rings and verge escapement. The movement is stamped DFDB and numbered K2002, possibly Denis Francois Dubois, who was received Master in 1767. The base and its protective dome are not original. Photo courtesy Sothebys Geneva.

Figs. 4a, b, c. An attractive and interesting French Skeleton clock by the well known maker Le Gros, still with its original fire gilding to the attractive scroll base which is decorated with flowers and leaves.

The two train movement has silk suspension to the sunburst pendulum and external count wheel strike on a bell. The dials are beautifully executed. The convex upper enamelled dial has Arabic numerals for the hours; minutes marked every 15 minutes and a gilt star every 5 minutes. It is signed Le Gros horloger de la Reine (Paris).

The clock is fitted with full perpetual calendar work i.e. it even corrects for leap year, the lower dial having three centre sweep hands which indicate from within out a) The day of the week b) The month of the year and the Zodiac c) The day of the month. There are gilt stars between the days of the week and the months and the zodiacal signs and the dates are written in burgundy. Most interestingly the back of the dial is signed by the well known enameller Barbichon and dated October 1785. Height: 11¾'' (30cms.)

Figs. 5 & 6. Two interesting clocks both made by Jean Louis Bouchet, Paris 1762-1789, who was clockmaker to the King. The one shown on the left gives the moon phases above the main dial and days of the week and days of the month below whereas the one on the right shows both the days of the week with their dieties and days of the month on the bottom left hand dial and moon phases at bottom right. Both clocks have countwheel strike on a bell, silk suspended sunburst pendulum and anchor escapement. A further, more complex example of his work is seen in Figs. 7a and 7b. Photos courtesy Christies.

are the series of what might be termed "extreme skeletonized clocks (**Figs. 76-79**) and the "Great Wheel Clocks" which, most unusually for French work, usually employ a fuzee.

A weight driven skeleton clock which was made as a limited series, is that seen in Fig. 85. An unusual feature of this is the use of a dummy weight put there purely for appearance to balance the driving weight. Recently this particular design has been reproduced.

By the mid 19th century, the French had started to produce a limited range of attractive small skeleton clocks in relatively large numbers, one particular example being that which is now termed "The Great Exhibition Skeleton" because it was shown there in 1851 when it was believed some 10,000 were sold.

Around this time considerable ingenuity was being expended in producing clocks of long, often year, duration. These employed anything from one to nine mainsprings and made use of various ingenious designs of pendulum, suspension and escapement, all with the aim of reducing the power required to a minimum. An example of one of these is seen in Fig. 88.

The ingenuity displayed in the construction of many of the skeleton clocks made in France is so great that it is probably best just to let the clocks speak for themselves and describe each one in detail rather than discuss their technical merits at this stage. However, there is one feature, the remontoire, which was used by many of the best French makers, which is of such great technical interest that it is described and illustrated in a separate chapter.

Figs. 7a, b. A skeleton clock of the highest quality and considerable complexity signed Bouchet Horloger du Roy Paris, circa 1780. It has an inverted Y frame with hour and quarter striking trains mounted at the bottom of each leg and a large spring barrel for the going train, in the centre. This powers the Robin type remontoire, the decorative wheel of which may be seen to one side of the frame with its weight and counterweight at the bottom.

The main dial has a centre sweep seconds hand and shows both mean and solar time by means of the equation kidney seen towards the bottom of the close up. Immediately below the main dial is a year calendar ring with teeth on its' outer aspect which rotates against a vertical central pointer. The gridiron pendulum may just be seen to one side of the frame. Photos courtesy Stender B.V., Holland.

Figs. 8a, b. An unsigned French skeletonized mantel clock made for the English market with Bisque porcelain statuettes resting on an ormulu mounted white marble base. The main dial has center sweep seconds, minute and hour hands and the subsidiary dials show the days of the month and the months of the year.

The movement has an inverted Y frame; external dead beat escapement with a pendulum with a massive bob and count wheel strike on a bell. It has a going barrel for the strike train and the escapement is driven by a weight which is rewound once a minute by a remontoire taken off a separate spring barrel. Photo taken by courtesy Schoonhoven Museum, Holland.

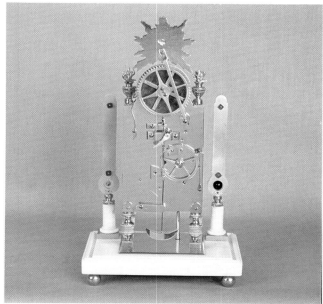

Figs. 9a, b, c. It is difficult to know whether to term this skeleton clock, which would have been constructed around 1780, English or French as although it is signed Martinet, London, presumably having been made by James Martinet's Family who emigrated from France in 1700, its' style is typical of French products of the Louis XVI period. The beautiful enamelled dials, as on this clock, were something we were unable to achieve in England and thus it is not surprising that the back of the dial plate is signed by a French maker, Delauney. Similarly the beautifully chased and finished ormolu mounts are of the finest quality and were probably also produced in France.

The clock rests on a white marble base with brass ball feet. Fixed to the front is an enamelled plaque on which is inscribed "Let Time Fly and Make Use of It" and surrounding this is a gilt bezel. Rising up on either side of the base are two turned marble columns with gilt capitals which carry thermometers. Both are attractively shaped and have scales made of engraved and silvered brass. That on the left employs spirit and is graduated -10C to +30C. Whilst that on the right is mercurial and graduated 0-70 Reamur. The fully exposed 2 train movement which strikes the hours and halves on a bell mounted within and below the base, rests on two ornamentally turned supports. It employs anchor escapement and silk suspension to the

pendulum which has a crescent shaped bob and a travelling clamp, a typical English feature.

The dial, a very fine part of the clock, has a shaped brass plate from the bottom of which is suspended a gilt swag of grapes with flaming urn finials on either side of the breakarch top. Enamelled plaques to the bottom left and right bear the inscription "Martinet" and "London". Above these is the main convex enamelled dial which has centre sweep minute, hour and date hands. There are gilt stars at the 5 minute marks and the Zodiaccal calendar is indicated with the appropriate signs.

A further three convex enamelled dials at the top give, from left to right, the day, the four seasons and the date. Above the dial plate is a beautifully executed lunar disc with the age of the moon being indicated around its' periphery. This is read off against a pointer attached to the Apollo mask surmounting the clock which is protected by a glass case with arched top and gold gesso to the edges.

Three similar clocks are known to exist. One is in a private collection in the UK, another is in the Nathan-Rupp collection in the Historischen Museum, Basel and a third, also signed Martinet, London, but having bisque figures on either side, is in the Musee des Arts et Metiers, Paris.

Fig. 10. A particularly interesting and complex skeleton clock which was made by Marc Bouchet, Horloger à l'Observatoire et Horloger du Roi, circa 1780-85. This clock is referred to in *Dictionnaire des Horlogers Français* by Tardy p. 70. The dials are signed by the emminent enameller Coteau. The quarter striking month duration movement has perpetual calendar and is fitted with a pin wheel escapement and remontoire. The horizonal rotating enamelled disc at the top indicates the years from 1800-1820, that on the left gives the zodiaccal signs with a 31 day calendar dial below it and on the one on the right the four seasons are depicted with the day of the week dial below.

The main dial shows seconds, minutes, hours and equation of time and the dial immediately below it is a year calendar. At the bottom left the age of the moon is indicated and to the right of this the times of sunrise and sunset. Photo courtesy G. Baumgartner, Switzerland.

Ferdinand Berthoud 1727-1807.

Figures 11, 12 and 13 show three fine clocks by this emminent maker. He was born in Switzerland in 1727 and moved to Paris some 20 years later where he rose to become one of the leading figures in French horology. He was in the forefront of chronometer development, designing his own form of detented escapement. His Horloge Marine No. 1 is illustrated on page 52 of Tardy's *Dictionnaire des Horlogers Français* where a much fuller account of his achievements is given.

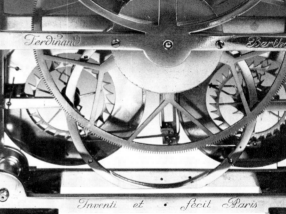

Figs. 11a & b. The main dial of this clock, which is signed on the backplate 'Ferdinand Berthoud Inventi et Fecit, Paris,' has center sweep seconds, minute and hour hands. To the bottom left days of the month are shown on the outer edge of the chapter ring, on its' inner and upper aspect days of the week and below the relevant dieties. The upper half of the dial on the right gives the months of the year and the lower half the appropriate zodiacal sign. Photo courtesy Stender B.V. Holland.

Fig. 11c. Skeletonised table regulator by Ferdinand Berthoud, Paris, circa 1785. The main dial has center sweep seconds, minute and hour hands and shows both mean time (out time) and solar time. In the center is a glass plate signed in white 'Ferdinand Berthoud Invenit et Fecit' and through this may be seen the engraved and silvered cam for the equation work.

The dial beneath the main dial has two gilt hands for the year calender which gives the days of the month and the Zodiac. The dial to the bottom left gives days of the week, and that to the right the age of the moon. The calender work is full perpetual, ie. it automatically corrects for the different lengths of the months and leap year. All the dials have delicate inner and outer gilt bezels.

The clock rests on a rouge marble base applied with beautiful fire gilt repousse work and on it rests a girl reading being offered garlands of flowers by two children.

The movement, which has a 9 rod gridiron pendulum, is of the highest quality. The thick plates are beautifully fretted out and the covers of the spring barrels are most attractively skeletonised. There is a large spring barrel for the going train, which automatically lifts the driving weight seen to the left of the clock every 30 seconds, and on the right between the two dials may be seen the mechanism for the remontoire. The clock chimes the quarter hours and strikes the hours on a bell. Photo courtesy Bobinet, London.

Figs. 12a, b, c. A month duration pillar clock signed on the backplate Ferdinand Berthoud, circa 1800. It has micrometer screw adjustment for the suspension of the gridiron pendulum mounted on the backplate. The dial has an engine turned center which makes it somewhat difficult to pick out the details of Berthoud's own escapement which is so prominently displayed in the lower half and thus it is illustrated here (Fig. 12c, next page). A very similar escapement by Berthoud is described by Tardy in *"French Clocks. The World Over"* Part Two 5th Edition p. 171. Norman Langmaid Collection, USA.

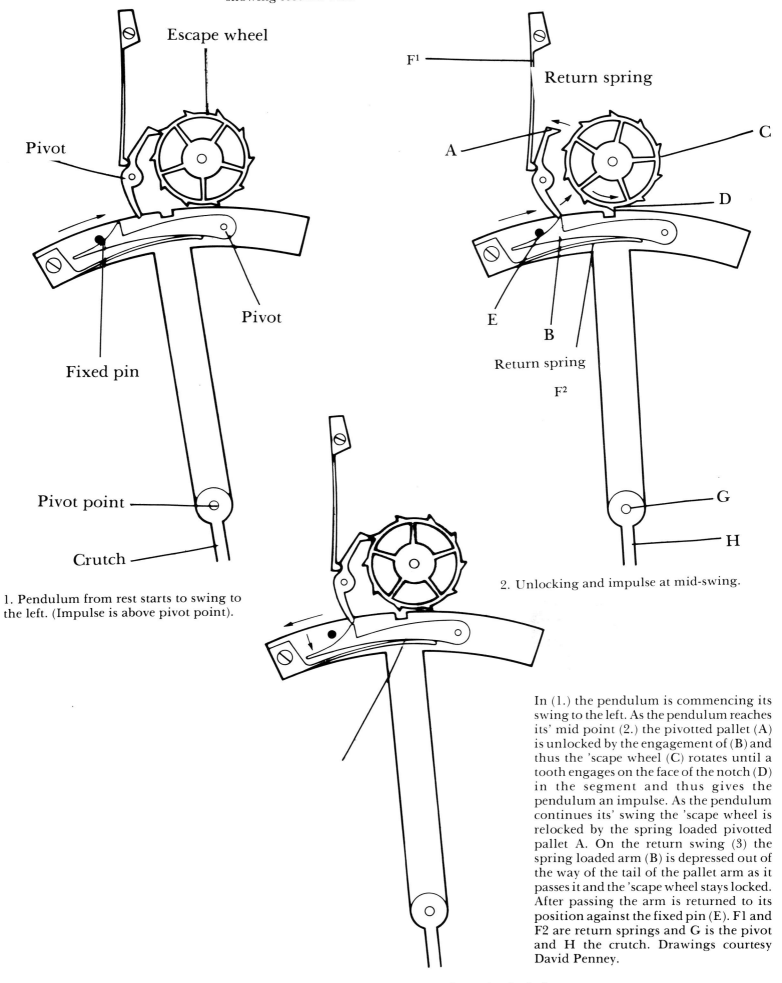

Fig. 12c. Coup perdu (lost beat) for 1/2 seconds pendulum showing seconds beat.

Escape wheel

Pivot

Pivot

Fixed pin

Pivot point

Crutch

1. Pendulum from rest starts to swing to the left. (Impulse is above pivot point).

F¹

Return spring

A

C

D

E

B

Return spring

F²

G

H

2. Unlocking and impulse at mid-swing.

3. In passing the escape wheel remains locked.

In (1.) the pendulum is commencing its swing to the left. As the pendulum reaches its' mid point (2.) the pivotted pallet (A) is unlocked by the engagement of (B) and thus the 'scape wheel (C) rotates until a tooth engages on the face of the notch (D) in the segment and thus gives the pendulum an impulse. As the pendulum continues its' swing the 'scape wheel is relocked by the spring loaded pivotted pallet A. On the return swing (3) the spring loaded arm (B) is depressed out of the way of the tail of the pallet arm as it passes it and the 'scape wheel stays locked. After passing the arm is returned to its position against the fixed pin (E). F1 and F2 are return springs and G is the pivot and H the crutch. Drawings courtesy David Penney.

Figs. 13a, b, c. A year duration skeletonised table regulator in the possession of the British Museum by whose courtesy these pictures were taken. It has double steel strip suspension to the gridiron pendulum, beat regulation, pin wheel escapement and a centre sweep seconds hand. Note the highly individual crossing out of some of the wheelwork. It is signed beneath the fine enamelled dial "Pendule allant un an Executee par Ferdinand Berthoud."

Fig. 14. A fine skeleton clock by Deschamps circa 1825. The main enamelled dial, with centre cut out to reveal the wheelwork, has 5 concentric hands, those for the hours and minutes being gilded and those for the seconds, weeks and months being of blued steel. Above the main dial the age and phases of the moon are displayed with a country scene. The frame has serpents winding up it on either side and there is a fine 9-rod gridiron pendulum with temperature indication. Photo courtesy Musée des Techniques. C.N.A.M., Paris.

Fig. 15. In figs. 15 and 16 are shown two skeleton clocks by Laurent á Paris which may truly be called fine works of art with their excellently designed arched frames, beautifully executed enamelled dials and royal blue and gilt enamelled panels and well chased and finished ormolu mounts and bezels. That seen here has pin-wheel escapement, hour and half hour strike on a bell and an enamelled chapter ring to the main dial which gives seconds, minutes, hours and days of the month. The subsidary dial at the top of the clock indicates the months of the year and that below the age and phases of the moon. Photo courtesy Christies, Geneva.

Figs. 16a, b, c. This clock is very similar to that already
described in 15 with the three dials giving exactly the same
indications: However in this instance the pin-wheel
escapement is controlled by a large balance wheel mounted
beneath the arch of the frame and behind the lunar dial.
Photo 16a courtesy Christies, Geneva. Photos 16b and 16c
courtesy private collection.

Figs. 17a, b. A skeleton clock formerly in the collection of Major Heathcote, the frame of which is very similar to that used by Berthoud for his year clock seen in Fig. 13. It has a center sweep seconds hand; pin wheel escapement and full calendarwork. It is somewhat unusual in that the two trains are mounted one above the other. All the strikework including the mainspring, countwheel and bell are at the bottom of the frame with winding taking place through the centre of the lower calendar dial whilst the winding square for the going train is immediately below 6 o'clock. The gridiron pendulum has steel strip suspension and beat regulation is provided.

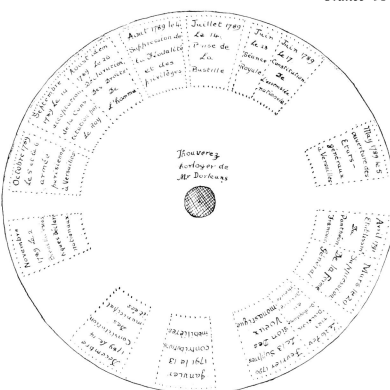

Fig. 18b. The calendar disc, copied actual size.

Figs. 18a, b. *Thouverez revolutionary skeleton clock.*
This clock which in overall concept and frame design bears a close resemblance to that of Berthoud seen in Fig. 11, has one additional very interesting feature, a calendar which records the fifteen principal events of the French revolution from May 1789 to April 1791. These are:

Janvier 1791 le 13
               Contributions Mobilieres.
Fevrier 1790 le 13
           Supression des Voeux Monastiques.
Fevrier 1790 le 26
         Division du Royaume en Departments.
Mars 1790 le 20
          Supression de la Ferme Generale.
Avril 1791 le 4
          Establissment du Pantheon Francais.
Mai 1789 le 5
       Ouverture des Etats Generaux a Versailles.

Juin 1789 de 17
         Constitution de L'Assemblee Nationale.
Juin 1789 le 23
               Seance Royale.
Juillet 1789 le 14
              Prise de la Bastille.
Auot 1789 le 4
Supresssion de le Feodalite et des Privileges.
Aout 1789 le 26
          Declaration des Droits de L'Homme.
Septembre 1789 le 14
        Acceptation de la Constitution par le Roy.
Octobre 1789 le 5 et le 6
         Armee Parisienne a Versailles.
Novembre 1789 le 2
        Biens Ecclesiastiques Declares Nationaux.
Decembre 1789 le 14
         Constitution des Municipalitees.

The calendar disc written in freehand and seen in the aperture below 6 o'clock, is shown full size in Fig. 10b. It moves some 15 times each year.

The clock was made in 1790 by Louis Thouverez of Paris for the Duke of Orleans who voted for the death of his cousin Louis XVI and subsequently followed him to the guillotine in 1793. His eldest son Louis Philippe became King of France in 1830. It is possible that when he was deposed and exiled to England that he brought the clock with him.

It is signed at the top of the chapter ring *Thouverez Hger De Mr D'Orleans* and at the top of the calendar mask *Inventor Fecit Thouverez.* Inscribed across the bottom is:*Le Tems a Pris un Corps et Marche Sous Nos Yeux.* Formerly in the Collection of Major Heathcote.

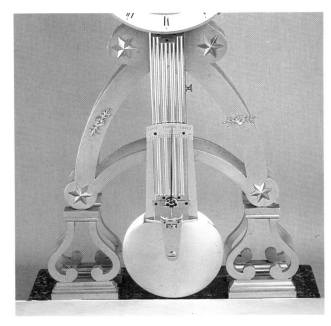

Figs. 19a, b, c. Fine French Quarter chiming skeletonised table regulator circa 1800. The substantial movement some 6'' in diameter, which chimes the quarters on two bells, has a large external count wheel and the pin wheel escapement is also mounted on the backplate. The narrow enamelled chapter ring has Roman numerals and is fitted with delicate inner and outer engine turned gilded bezels; there is a centre sweep blued steel seconds hand and the hours and minutes are of gilded brass.

The beautifully executed 9 rod gridiron pendulum is a central feature of the clock, being mounted in front of the front frame. A long hand rises from the base of the gridiron to give temperature indication against a silvered brass scale. The gilded arched frame is supported by scrollwork which rests on a flecked green marble base with gilded feet. Height: 21½'' (54.5cms.)

Fig. 20. A fine Louis XVI weight driven skeleton clock signed by Charles Bertrand, Clockmaker to the Academie Royale des Sciences (C.N.A.M.)

The movements driving weight, with engine turned decoration, may be seen suspended from an arm to the left of the dial. The escapement is controlled by a large balance wheel mounted horizontally between the bottom of the four pillars and is protected by guards top and bottom. The central arbor may be seen running up from this to the movement.

All the dials are enamelled and have inner and outer gilt bezels. The main dial has a centre sweep seconds hand and an attractively counterbalanced minute hand. The dial to the bottom left gives the day of the week and the ruling diety and that on the right the day of the month (0-31). Photo courtesy Musée des Techniques C.N.A.M., Paris.

Figs. 21a, b. An extremely impressive French skeleton clock, made around 1800, formerly in the collection of Major Heathcote. Although it is relatively complex this impression is enhanced by the delicate and extensive fretting out of the frames in the form of stars, scrolls and circles. Indeed when looking at the clock it is initially by no means easy to differentiate between the wheelwork and the frames.

It has a single massive mainspring with skeletonised barrel which powers the striking train directly and the going train indirectly by means of a spring remontoire which it rewinds during striking.

The pin wheel escapement has very long pallet arms and is controlled by a pendulum. This has knife edge suspension with screw regulation for fast/slow immediately above it and is provided with beat regulation.

The very delicate count wheel for the hour and half hour strike on a bell concealed in the base may just be discerned against the maze of other wheels to the centre right of the frame.

The wheel to the top right of the clock has engraved on its rim a 30 day calendar and in the centre is a semi-circular dial calibrated 20-0-25 which is inscribed on the left Humidite (Dampness) and the right Secheresse (Dryness). Photo 21a Courtesy Keith Banham. Photo 21b Courtesy Mr. Heathcote.

Fig. 22 *A fine and complex musical clock signed J. Van Hoof et Fils à Anvers 1796*

This fully skeletonised musical clock has 3 dials. That at the top has a rotating globe moon in the centre and outside this a rotating double 12 hour chapter ring which gives the time at any of the 12 cities in the world marked around the periphery of the dial. The main dial has gilt hour and minute hands and a blued steel centre sweep seconds hand and the lower dial gives the days of the week and month and the month of the year.

At the top of the clock are two bells for the hours and quarters and at each hour the tune "Roque à Paris" is played on two basically independent mechanisms both of which have large spring barrels and fuzees. One has a brass pin barrel and employs 11 bells and hammers and the organ at the bottom uses a large wooden pin barrel in conjunction with 20 flutes mounted vertically across the back of the movement; some of which may be seen on the left in the photo. Photo courtesy Musée des Techniques C.N.A.M., Paris.

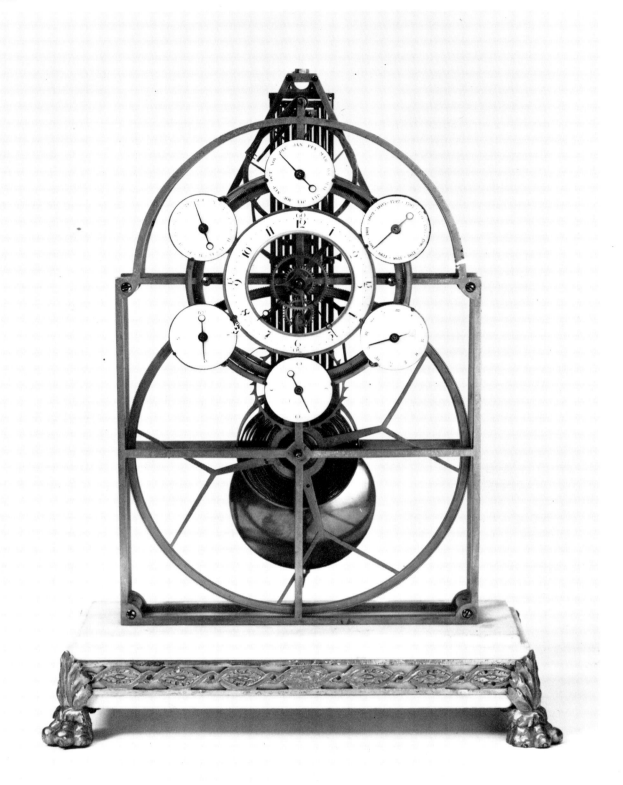

Fig. 23. A fairly small unsigned French late 18th century
perpetual calendar skeleton clock of great originality. It is
fully skeletonised, even the barrel cover being omitted. A
pin wheel escapement is employed with a nine rod
gridiron pendulum having knife edge suspension and
beat regulation.

Arranged radially around the main dial are 6 subsidiary
dials which, starting at 12 o'clock and proceeding clockwise indicate 1) Months of the year, 2) The years from 1792
to 1803, 3) The months with their zodiacal signs, 4) The
phases of the moon, 5) The permanent calendar for 4 years
including leap year, 6) Days of the month. Height: 14''
(36cms.) Photo courtesy Ineichen, Zurich.

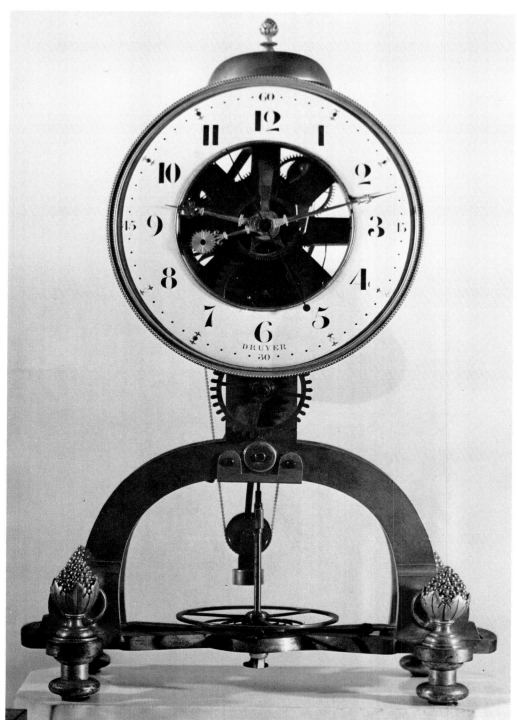

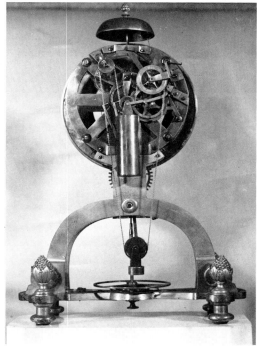

Figs. 24a, b. An interesting skeleton clock with several unusual features. It has a fairly large spring barrel which may be wound either by using its' own arbor with the attendant risk of damage to the center sweep seconds hand, or by means of a winding square placed below 6 o'clock which is connected to the single large spring barrel by an intermediate wheel. A remontoire is employed for the going train the weight and counterweight of which are shown in the rear view and the striking train is spring driven.

The pin wheel escapement, the inverted pallets of which may be seen to the left of the driving weight, is controlled by a large plain balance wheel mounted at the base of the frame with its' hairspring immediately below it. The enamelled dial is signed by Druyer who was born in 1755, apprenticed 1769 became Master in 1786 and continued work in Paris until 1810. Collection Dr. S.P. Lehv, U.S.A.

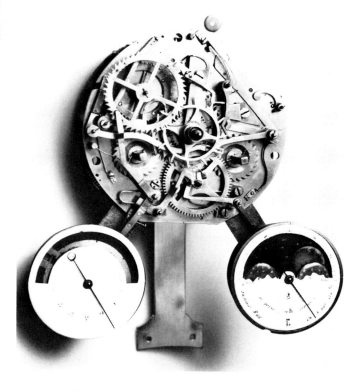

Figs. 25a, b. A rare and interesting late 18th century French skeleton clock with full perpetual calender and astronomical indications, signed on the white enamelled dial Jul. Pre. Mel. Bardonil. Inside the Roman numerals for the hours is a 0-31 day calender and below 12 o'clock an aperture with revolving silvered brass ring which indicates the months with their number of days; the signs of the Zodiac and the times of sunrise and sunset. A small aperture above 12 o'clock gives the year.

The dial on the bottom left shows the days of the week with their diety and a silvered brass ring displayed in the aperture against a cobalt blue enamelled background gives the months. In a square aperture which may just be discerned in the 10 o'clock position the Dominical letter is displayed. In the ecclesiastical calender the first seven letters of the alphabet are put in front of the day of the week. Which letter is put in front of each day is determined by the first seven days of January. Thus if January first is a Friday then Friday will have the Dominical letter A throughout the year, Saturday B and so on.

The enamelled dial to the bottom right gives the age and phases of the moon.

The substantial 2-train movement has countwheel strike on the hours and the halves and employs an anchor escapement with half seconds beating gridiron pendulum with steel strip suspension. There is a special train for the perpetual calendar work which is most unusual. There are two cams set on the same arbor which revolves every 4 years. The lower arm allows the date to pass automatically for the 28, 30 and 31 day months; the other operates the months, the year, and every four years allows February 29th to be set automatically. The frame of the clock is gilt bronze and it is set on a flecked grey-white marble base. Height: 19'' (48cm). Photo courtesy Jean-Claude Sabrier, Paris.

Fig. 26. This attractive style seems to have been introduced in France in the last quarter of the 18th century. The fire gilt cases are always of high quality and frequently decorated with enamelwork as in this case. The movements are usually skeletonised and often incorporate interesting mechanical features.

The clock illustrated here has centre sweep minute, hour and date hands; shows moon phases above the main dial and indicates days of the week in the centre of the beautifully executed pendulum bob. A similar clock is shown in the chapter on escapements. Photo courtesy Ineichen, Zurich.

Fig. 27. Late 18th century enamel mounted ormolu mantel clock with royal blue surround to the chapter ring, signed on a decorative plaque below this 'Faisana à Paris'. The skeletonized movement has anchor escapement and external count wheel strike on a bell. Height: 1' 6'' (45.5 cms.)

Fig. 28. A late 18th century French skeleton clock with fretted out plates and particularly fine wheelwork with "star" crossings. Photo courtesy Stender B.V., Holland.

Fig. 29. This clock bears a marked similarity to those made by Bouchet seen in Figs. 5 & 6 and that shown in Tardy Part 2, 5th edition, page 241. However these employ a pendulum whereas the pin wheel escapement of this particular clock is controlled by a large balance mounted on the baseplate. It is signed between the dials "Retablie par Charpentier 1805". At the top is shown moon phases, to the bottom left the days of the week and their dieties and to the right the day of the month. Height: 16½" (42cms.) Photo courtesy Uto Auktions A.G., Zurich.

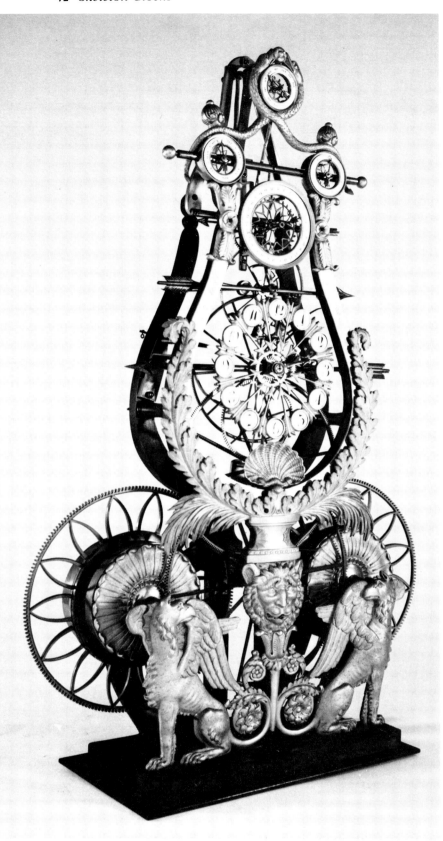

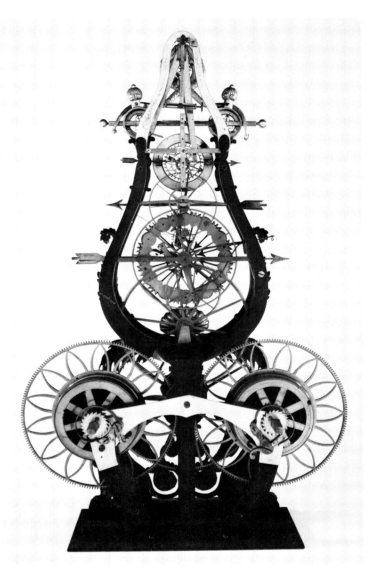

Figs. 30a, b. A fine Empire style skeleton clock of year duration decorated with griffins, a lion mask, arrows, a conche shell and two serpents entwined around the upper dials.

It has two skeletonised going barrels which are wound from the rear with both driving through two beautifully pierced out great wheels onto a single common central arbor. All the wheelwork including the star wheel is of great delicacy and varies throughout the train, each wheel being treated in a different way. It has a Debaufre escapement and is signed on the frame J.G. Aeits à Tongres.

The main dial has raised enamelled plaques for the numerals and the subsidiary dials at the top show, from top to bottom, months, days of the week, the dieties and days of the month. Height: 20½'' (52cms.) Collection Dr. S.P. Lehv.

Figs. 31 & 32. Two very similar skeleton clocks, one (31, left) signed J.F. Henri Motel à Paris at the bottom of the dial and in a segment at the top Horloger de la Marine Royale who probably made both clocks, and the other (32, right) signed W. Prinzlau Hamburg.

Both have vertically mounted verge escapements controlled by a balance wheel at the top of the clock. However whereas the Hamburg clock only has a relatively small seconds dial fretted out in the centre in the form of a star the Motel clock has a very large seconds ring almost dominating it. The two subsidiary dials are the same on both clocks those on the Motel clock bearing the inscriptions below them Ascension Diurne du Soleil (Sunrise and sunset) and Revolution et Phase de la Lune (Revolution and Phases of the Moon.)

Although the skeletonised equation kidney may be seen in the centre of the dial of each clock their layouts are different. Henri Motels clock is calibrated 1-31 on its' outer edge but has two sets of divisions, one giving the solar and the other mean solar days. On the inner aspect of the central ring is a year calendar and on the outside of the top half of this the "equation of time" is indicated.

The Hamburg clock differs from the Motel in having hands which show both solar and mean solar time from which of course the equation of time can be calculated, instead of an equation of time hand. Like the other clock it has a central year calendar. Fig. 31 Courtesy Stender B.V. Holland. Fig. 32 Courtesy Jean-Pierre Rochefort, Paris.

Fig. 33. An early 19th century musical skeleton clock
signed Sagniez a Montreuil. It has ting-tang quarter
chime on two bells and strikes the hours simultaneously
on the same two bells employing a large count wheel
which may just be seen through the centre of the dial.

One of ten tunes is played at the hour using 13 bells and
25 hammers and there is automatic change to a different
tune at every hour. The linkages running up on either side
of the clock outside the large great wheels are for tune
change and stop/start. Because of the lack of space the
large 'scape wheel has been offset to one side of the frame
and part of it may be seen to the top right of the chapter
ring. The pallets are directly linked to the centrally
suspended pendulum by a rod. Height: 20¼''(51.4cms.)

Figs. 34a, b, c, d, e. Although it is probably fair to say that most of the very fine French work was done in the period 1760-1830, many superb clocks of great technical interest were produced in France well after this era by clockmakers such as the Lepautes and the Houses of Breguet, Le Roy, Lepine and Pons to name but a few. In many cases these followed on in the idiom of earlier clocks, for instance an appreciable number of excellent longcase regulators of classical proportions were produced throughout much of the 19th century.

A clock which typifies the fine work being done in France, quite frequently in the country, in this later period is this beautiful multidialled skeletonised table regulator which at first sight might have been made around 1790 but is in fact signed on both the dial and the frame Eel Escoube, Toulouse and additionally dated on the frame Fecit 1854.

The Frame design, an inverted Y, is similar to that used from the 1770's onwards and both plates are very substantial. There is a large going barrel; pinwheel escapement and a 9 rod gridiron pendulum with knife edge suspension. The main dial has centre sweep seconds, minute and hour hands and an additional minute hand giving solar time and thus the equation of time. Immediately below the main dial is an engraved and silvered year wheel marked with the months of the year and behind this may be seen the equation cam.

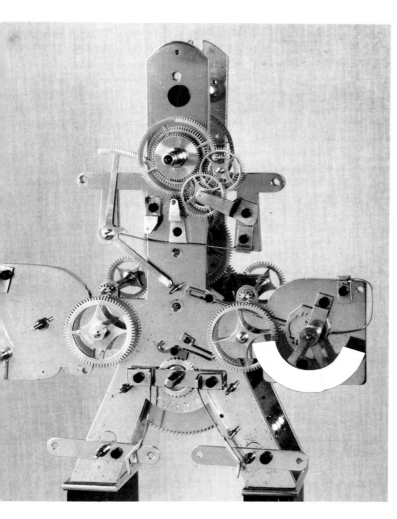

The top left dial gives the day of the week and that on the right the day of the month. The dial below and to the left gives the moons age and phases and on the right may be seen the sun rotating in a recess against a blue ground. White enamelled shutters move up and down throughout the year at each end of the recess so that the sun always rises and sets behind these at the correct time, which may be read off against a scale graduated III to XII twice. The mechanism for the shutters may be seen in the close up photo. Each shutter is fixed to an arbor, at the other end of which is a rack. These are activated by the spring loaded bar being moved in or out by an eccentrically mounted disc, not seen here, pressing on a pin at its' end.

The two dials at the bottom of the clock display pictorially the signs of the Zodiac and the Seasons and in the overall view with the dials removed the equation work may be seen.

The clock rests on two flecked green marble rectangular brass capped pillars which in turn rest on a matching octagonal base. The clock is protected by a glass dome standing on an oval base and is some 20'' (51 cm) high.

THE WORK OF VERNEUIL & NICOL

Verneuil appears to have specialised in fine skeleton clocks and all of those produced by him bear a close family resemblance although they vary appreciably in complexity. Some four are illustrated here and in addition we are including two which whilst unsigned may well have been made by him. Two of the clocks are just signed Verneuil and have no place name but a third bears Paris on the dial and thus it seems reasonable to assume that they were made by the Verneuil who was working at 42 Rue St. Honore in Paris, although it could just possibly have been his son who was at Rue du Contrat Social in 1806 and Fg St. Martin in 1815.

In addition to Verneuils signature, which is usually engraved up the side of the front plate but is sometimes seen on the dial (Fig. 41) the name Nicol à Paris also appears on these clocks, sometimes on the backplate and in one instance on the dial (Fig. 38); however no such maker is recorded by Tardy. It may be that Nicol supplied the plates or even the Blanc-Roulants (the basic unfinished movement) for the clocks or possibly vice-versa. Alternatively they may, like Parker and Pace in England, have worked together.

What is particularly interesting is that the shape of the plates which are mostly quite similar and can be seen best in Fig. 37 bears a close resemblance to those used by Berthoud (Fig. 13) and the ones employed on the clock seen in Fig. 17. Thus it may be that either Nicol or possibly Verneuil supplied plates or even Blancs-Roulants to other makers.

Besides those signed by Verneuil and the anonymous clocks we are including two which bear the signature of Le Roy as these are very similar in their overall concept and execution.

All the 8 clocks in this section except one incorporate calendarwork, mostly laid out in a similar way; some have enamelled and some silvered brass dials and seven of the eight clocks rest on figured marble bases. A gridiron pendulum is employed on all of the clocks except that seen in Fig. 36 and all would have been made in the Empire Period (1800-1830).

Two further examples of Verneuils work may be seen in the book on the Salomons Collection[1].

---

[1] Daniels, G. & O. Markarian. *Watches & Clocks in the David Salomons Collection.* Sotheby, 1983.

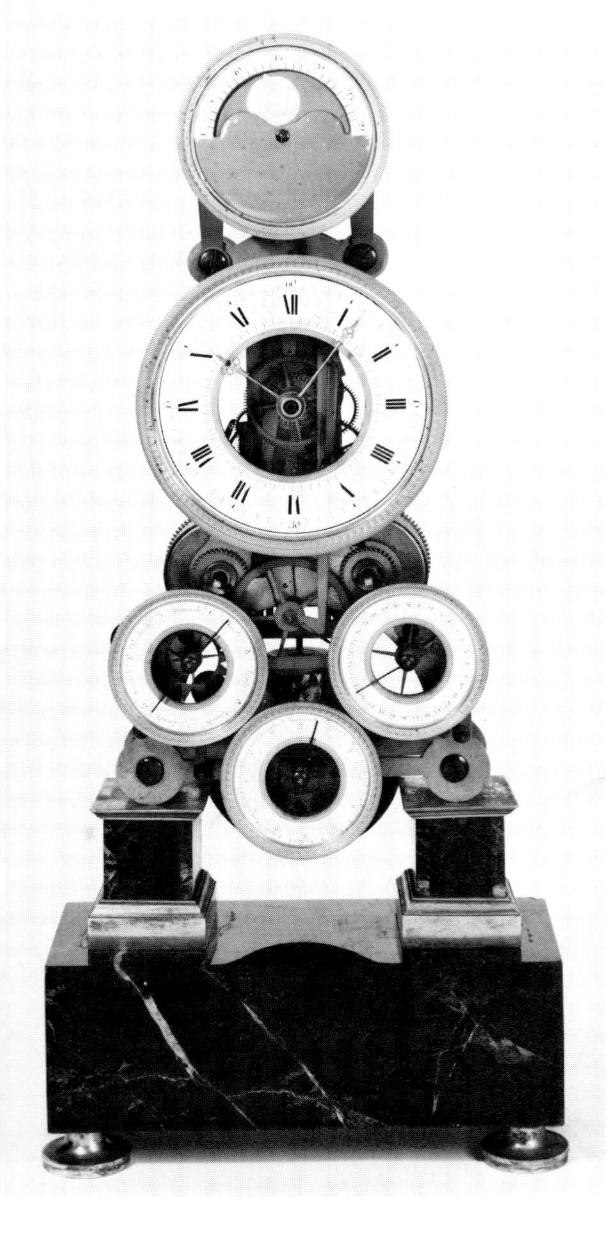

Fig. 35. A French early 19th century skeleton clock signed Le Roy à Paris. The two train movement has count wheel strike on a bell, pin wheel escapement, gridiron pendulum and attractively executed enamelled dials showing moon-phases, the day with its' diety; the days of the month and the months of the year with the relevant zodiacal sign. Height: 21¾'' (55 cms.) Photo courtesy Phillips.

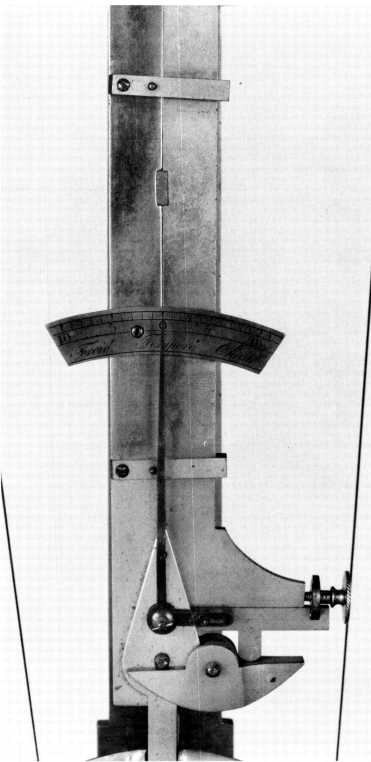

Fig. 36i, ii. A month duration skeleton clock signed B.C. Le Roy à Paris with the lines from both driving weights, which fall into recesses in the base so as to give the required duration, winding onto a single central barrel. It has a year calendar ring which rotates against a fixed pointer which may just be seen at the bottom. Pin wheel escapement is employed and the compensated pendulum is to Berthouds' design. The principle is similar to that of Ellicotts but uses only one pair of bars. As one expands down it presses on one end of a pivotted lever and thus automatically raises the bob attached to the other end and keeps the effective length of the pendulum constant. A pointer attached to the pivotted lever indicates the degree of thermal compensation being achieved. Height: 25'' (63.5 cms.) Norman Langmaid Collection. U.S.A.

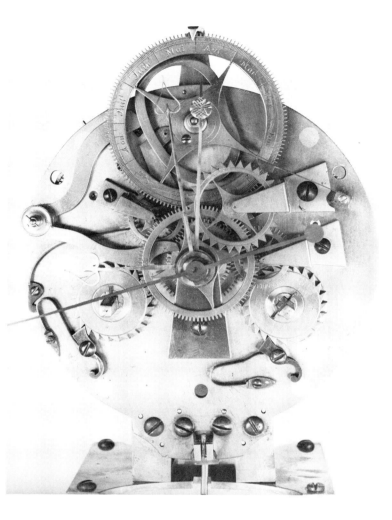

Figs. 36a, b, c. A most unusual two train early 19th century French skeleton clock of 30 days duration with count wheel strike on a bell; pin wheel escapement with inverted pallets and a nine rod gridiron pendulum. Interestingly the 'scape wheel, including all the pins, which are flattened, has been taken out of one piece of solid brass. The year calendar disc may be seen in Fig. 36b and the top of this, together with its' pointer is visible just above the silvered brass chapter ring at 12 o'clock. In the centre of the date ring is the equation kidney which enables the clock to show both mean and solar time. Height: 17¼'' (43.8 cms.) Albert Odmark Collection. U.S.A.

Figs. 37a, b. Early 19th century two train skeleton clock signed on the left side of the front plate "Verneuil" and on the backplate Nicol à Paris. It has countwheel strike on a bell; pin wheel escapement; 4- and 5-spoke wheelwork and a gridiron pendulum.

The centres of all the silvered brass dials have fine engine turning. The double hand on the left hand dial gives both the day of the week and the diety; on the right is a calendar dial graduated 1-31 twice and at the bottom a lunar dial. Height: 24½" (62.2 cms.)

Figs. 38a, b. A fine French skeleton clock of three months duration signed on the silvered brass dial Nicol à Paris and on the edge of the front frame by Verneuil. It has a centre sweep seconds hand and a pin wheel escapement with a nine rod gridiron pendulum mounted above it which has point suspension and shows the degree of temperature compensation being achieved by means of a pointer and scale mounted on its' front face.

The movement has two mainsprings for both the going and the striking trains; the former being wound through the main dial and the latter above the lunar dial.

The principal dial has in effect three rings; the outer one which gives the hours, minutes and seconds; an intermediate one on which the names of 53 cities are engraved, thus giving you their relative times, and in the centre there is a rotating crossed out disc calibrated 1-12 twice, once in Roman and the other Arabic, which gives you the actual times in these places.

The subsidiary dials on the left give the days of the week, their dieties and the days of the month and on the right are shown the months of the year and the four seasons. In the centre is a lunar calendar. Langmaid Collection. U.S.A.

Figs. 39a, b. A swinging clock signed on the narrow enamelled chapter ring Huguenin à Paris circa 1820. As can be seen the dial and fully skeletonised movement in effect form the bob of the nine rod gridiron pendulum which has knife edge suspension. Pin wheel escapement is employed with an inverted crutch which embodies beat regulation engaging a pin mounted half way up the vertical pillar supporting the clock. Height: 18'' (45.8 cms.) Photographed by Courtesy of J. Fanelli, New York.

Figs. 40a, b. A month duration French skeleton clock signed on the silvered brass dial Klaftenburger 157 Regent Street (the distributor) who was active from 1863-81. It is also stamped on the reverse of the dial by Raingo, presumably the maker. It has two going barrels with count wheel strike on a bell and Graham type dead beat escapement. A slightly unusual feature of the nine rod gridiron pendulum, which has knife edge suspension, is that there is an aperture in the cross bar immediately above the bob through which the top half of the graduated rating nut may be seen.

A month calendar is provided above the main dial and an unconventional feature is the long linkage which connects this to the dial in the base which indicates the days of the week. Height: 27'' (68.5 cms.). Formerly in the Heathcote Collection and now in that of Albert Odmark, U.S.A.

Figs. 41, a, b, c. This spectacular clock, signed by Verneuil, with strong Egyptian and Roman influence, stimulated no doubt by Napoleons famous victories around that time, was photographed at Schoonhoven in Holland by courtesy of the Museum authorities there. The two train movement has a pin wheel escapement and a beautifully executed nine rod gridiron pendulum with beat regulation.

The lunar dial at the top is one of the finest the author has seen on a clock. A lake with a boat and a castle is shown in the foreground and an erupting volcano behind. Pearls are set in at the bottom of each side of the lunar calendar. The dials to the bottom left give the days of the week and their dieties and the months of the year and on the right are shown the days of the month and the signs of the zodiac. Height: 30½'' (77.5 cms.)

Fig. 42. This clock is like that shown in Figs. 38a, b in that it has the pendulum mounted above it and is of three months duration with two going barrels but is only a timepiece.

It has pin wheel escapement with the wheel, only 3/4″ in diameter, lying immediately behind the seconds ring. It is signed on the enamelled chapter ring Verneuil, Horologer, Mechanicien à Dyon.

Fig. 43. This clock is interesting in that it is the only one of the series which is signed Verneuil Jeune à Paris and was thus probably made a little after the others. It has centre sweep seconds, minute, hour and 30 day date hands. The beautifully executed skeletonised wheel for the latter may be seen in the centre of the dial.

The large 9 rod gridiron pendulum is mounted at the front of the clock and embraces the cannon pinion. There is a cranked crutch with pinwheel escapement mounted on the backplate and outside countwheel strike on a bell. Height: 16½″ (42 cm). Photo courtesy Patric Capon.

Figs. 44a, b. A most interesting unsigned *year duration* skeleton timepiece with four going barrels and four great wheels, two working on either side in tandem, and driving onto a common central arbor. It has a substantial scroll frame with a duplex escapement mounted above it. The main dial is of English Regulator layout with a centre sweep minute hand and subsidiaries for the seconds and hours. In addition there are apertures through which are displayed days of the week and month and weeks and months of the year.

The lower World time dial gives the time in seventy different places which are inscribed on the central disc which rotates against the fixed 24 hour chapter ring. Height: 17½'' (44.5 cms.) Collection Dr. S.P. Lehv, New York.

Fig. 45. Another clock, also of three months duration, with a gridiron pendulum mounted above it, pin wheel escapement situated again immediately behind the seconds ring and count wheel strike on a bell. The lower part of the train employs a high count steel pinion and the arbors of all the wheels, which are 4 spoke, are carried on individual cocks screwed to the frames.

The pendulum has steel strip suspension with a rating nut above this and another rating nut next to the bob. The main dial is enamelled and all the subsidiary dials are of silvered brass with apertures, that at the top showing moon phases and marked Croissant (Waxing) and Decroissant (Waning). The bottom left hand dial shows the days of the month and on the right the months are indicated. The clock rests on a multicoloured marble base. Height: 26'' (66 cms.) Collection A. Odmark, U.S.A.

Figs. 46a, b. A fine quality unsigned skeleton clock, possibly by Verneuil, with a beautifully executed enamelled dial by Coteau, on which are shown seconds, minutes, hours and date, the star wheel for the latter being delicately fretted out. The rear view (Figs. 122a, b, c) shows the large count wheel for the strike, the pin wheel escapement and the remontoire. An attractive and relatively unusual feature is the mounting of the pendulum at the front of the clock. The lower subsidiary dials indicate the day of the month with the relevant diety and the months of the year with their zodiacal sign.

Fig. 47. French skeletonized table regulator of 21 days duration with coup-perdu pin wheel escapement enabling seconds to be indicated by the centre sweep hand even though a half seconds pendulum is employed. It strikes the hours and half hours on a bell. An attractive feature of this clock, which is of classical simplicity, is the beat plate resting on the black slate base. It is signed on the white enamelled dial Aldre Chambel, Paris. Height: 21" (53.5 cms.)

Having given some idea of the range and complexity of the French skeleton clocks which were produced, mainly on a one off basis from the mid-18th to the mid-19th century we will now move on to the skeleton clocks which were produced to similar designs usually in limited but sometimes in appreciable numbers. These are:

1. Fine skeletonised table regulators.
2. Great Wheel Skeleton Clocks, some of which employed a pendulum and others a balance.
3. Glass plated skeleton clocks.
4. Keyhole framed or extreme skeletonised clocks.
5. Portico or 4 pillar skeleton clocks.
6. The "Eagle" clocks.
7. Year Clocks.
8. The "Great Exhibition" clocks.

FINE SKELETONISED TABLE REGULATORS.

In the late 18th-early 19th century a beautiful series of skeletonised table regulators was produced of classic simplicity and all of fine quality. They bore the names of Lepine, Breguet, Robin and in one instance Valogne but their likeness is so striking that it seems probable that they were all conceived and also probably executed by one maker. Some of the clocks are weight driven whilst others employ a remontoire; some are striking and others not and the choice of escapement and pendulum varied. However all use a centre sweep seconds hand, they usually had levelling screws at the base and the frame design is very similar, being either with or without a notch on the arched base.

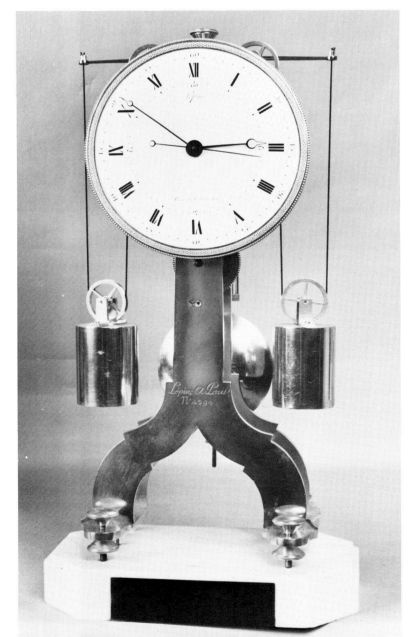

Figs. 48a, b. A fine weight driven skeleton timepiece with the lines from the two weights feeding onto a common central barrel, signed on the enamelled dial Lepine, Place des Victoires, Paris and on the front frame Lepine à Paris No. 4394. Dead beat escapement is employed with relatively long teeth on the 'scape wheel and thin adjustable pallets. The nine rod gridiron pendulum rests on a knife edge and beat regulation and maintaining power are provided.

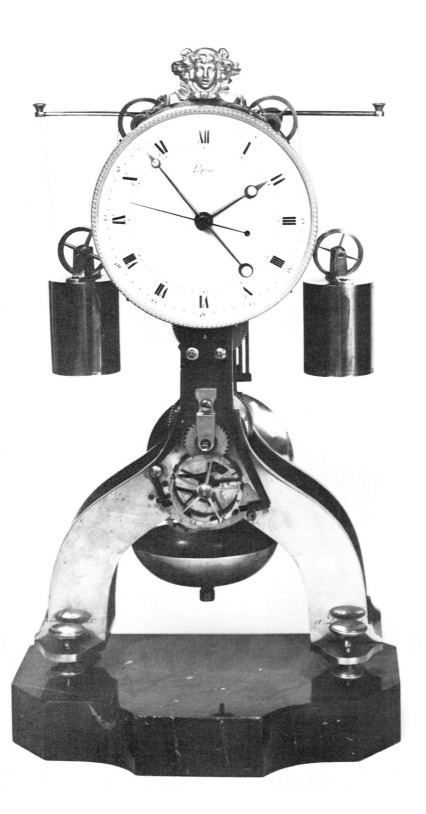

Figs. 49a, b. A similar clock to that seen in Fig. 48 with two driving weights to the going train, dead beat escapement, gridiron pendulum and maintaining power. In addition there is a spring driven striking train with the count wheel mounted on the front of the frame and the bell immediately below it. The mount placed above the chapter ring is similar to those seen on the clocks signed by Breguet. It is signed Lepine à Paris and numbered 4553. Photos courtesy Jean-Claude Sabrier, Paris.

Figs. 50a, b. A weight driven skeleton timepiece possibly a little later than the others in this series. Again it has the lines from the two driving weights feeding onto a central barrel and the front plate is identical to that seen in Figs. 48 and 51, however the bottom half of the back plate has been omitted, an engraved and silvered brass dial is used and the pendulum has a steel rod with a very large brass bob. A nice touch is the way in which the sinks for the pivot holes have been cut progressively deeper into the plates the higher up the train they are so as to reduce the bearing area of the pivots and thus the friction to a minimum. A pin wheel escapement is used with beat regulation. It is signed at the base of the backplate Valogne à Paris. Overall height 24½'' (62 cms.)

Fig. 51. The specification of this clock is virtually identical with the one signed by Lepine seen in Fig. 49. However it bears the name Robin à Paris. Minor differences are the tiered weights and the notches on the inner and outer edges of the base. The hour and minute hands would seem to be out of keeping with the rest of the clock. Height: 26⅜'' (67 cms.) Photo courtesy Luis Montanes, Soilo Ruiz-Mateus S/A.Jerez de la Frontera, Spain.

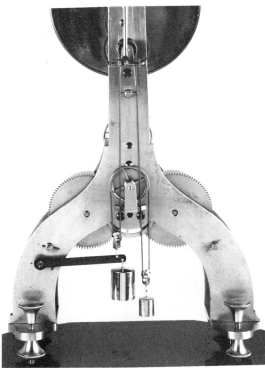

Fig. 52b.

Breguet

The number of clocks produced by Breguet was quite limited by comparison with his watches and it is likely that many of these were made outside his workshops, either being commissioned by him and made to his own designs or being fine pieces which were offered to him by other makers. On some occasions he would merely sign the piece but in other instances even though he had not made the clocks he both signed and numbered them.

George Daniels in his excellent book on ''The Art of Breguet'' does discuss this in some detail and goes to considerable lengths to explain Breguets attitude to his work and his business relationship with other clockmakers. Suffice it to say here that when one compares the two clocks signed by Breguet shown in Figs. 52 and 119 with those of Lepine illustrated at the beginning of this section one is left in little doubt that the basic design is that of one man, probably Lepine. However both the clocks in Figs. 52 and 119 are fitted with a remontoire and it may well be that Breguet specified this and possibly even designed it.

Fig. 52a.

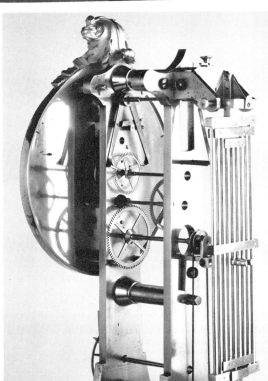

Fig. 52c.

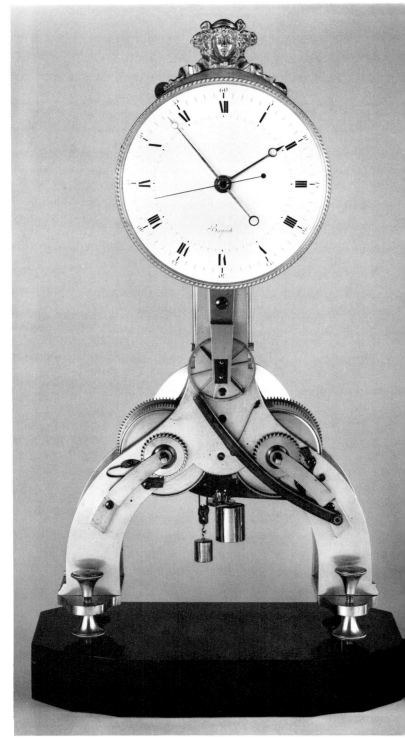

Figs. 52a, b, c. Skeleton clock, signed by Breguet, which is very similar to that shown in Fig. 119. In the views shown here the action of the remontoire may be seen quite clearly. In Fig. 52c the cord passing over the pulley mounted on the backplate is shown supplying power to the bottom end of the train. In Fig. 52b this same cord is seen passing round a small pulley attached to a lever at the end of which is the driving weight. As the weight and thus the lever descends another lever on the frontplate, which is mounted on the same arbor, also descends and in due course unlocks the remontoire wheel which releases the train and rewinds the driving weight. The small weight is only a counterweight which is used for keeping the cord tight and preventing it slipping on the pulley. Photos courtesy Jean Claude Sabrier, Paris.

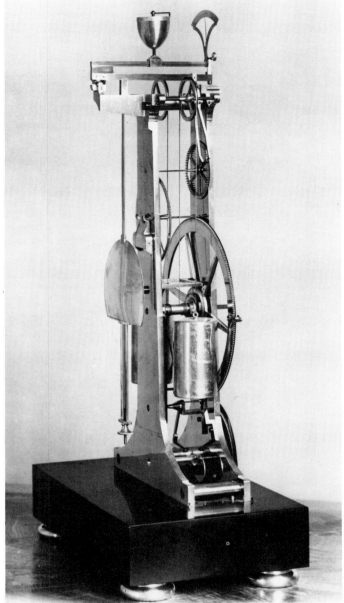

Figs. 53a, b. This shows one of the *pendule a trois roues* no. 3583 (3-wheeled clocks) made by Breguet between 1795 and 1818 of which four are known by George Daniels to still exist. A further two, made by Daniels himself in 1969 when he was London agent for Breguet, were certified by Mr. George Brown, proprietor of Breguets at that time and issued with certificates and the numbers 3224 and 3225. (Fig. 213b). Interestingly in his book George Daniels states that there is documentary evidence that the three wheel clocks were made to Breguets design by Samuel Roy at Chaux-de-Fonds in Switzerland.

It has a pin wheel escapement with a half seconds pendulum supported on a knife-edge. Compensation is achieved by a bimetallic strip above with a scale and hand to indicate any change of temperature. There are two driving weights the lines from which feed onto a central arbor via two pulleys mounted at the top of the frame. Attached to the central arbor is a 24 hour disc subdivided into 10 minute intervals. The indicator at the left with additional indicator above for equation of time. The frontplate is engraved with the days of the week which are indicated by the height of the tops of the weights. The dial at the bottom is engraved with the Julian and Revolutionary calendars. Photos courtesy George Daniels.

GREAT WHEEL SKELETON CLOCKS

This interesting series of clocks were nearly all made in the first 30-35 years of the 19th century and had many features in common. Thus they all used an inverted Y frame which in some instances was left flat as in figures 62 and 65a, b and in other cases was attractively shaped and grooved. The great wheel, the principal feature of the clock, was always 1½ to 2 times as large as the chapter ring and the intermediate wheel a little larger than the dial so that, in effect, it appeared to run around it.

All the clocks had a large 'scape wheel clearly visible above the dial and in the majority of cases a pendulum was used in conjunction with a tic-tac escapement with the pallets only spanning three teeth. However in some instances a pin-wheel escapement was favoured and where this was the case an anchor was often used for the pallet instead of the arms usually seen in conjunction with this form of escapement. Occasionally the 'scape wheel was controlled by a large balance wheel mounted level with and to one side of it.

An appreciable number of these clocks were fitted with a chain fuzee and in these instances the spring barrel, which was usually at the centre of the great wheel, was moved and either let into (Fig. 54) or mounted on (Fig. 60) the base. The treatment of the barrel cover varied; sometimes they were left solid, sometimes fretted out and occasionally left open.

The appeal of these clocks lies very much in their simplicity, but occasionally mounts were added and it is not uncommon to see plaques, usually with a classical motif, applied to the base.

An interesting technical feature sometimes used is Wolfs tooth wheel-work, a close up of which may be seen in Fig. 56c.

The majority of these Great wheel clocks are unsigned, however their similarity is so great that one cannot help feeling that they were all either made by one or a small group of manufacturers.

Fig. 54. A month duration chain fuzee skeleton timepiece with the spring barrel let into the white marble base. The movement which has only three pillars, has a Great Wheel 6¾'' (172 mm) in diameter with 385 teeth; the centre wheel is 4'' with 360 teeth and the 'scape wheel 2½'' (60 mm) with 78 teeth and uses an eight leaf pinion. The tic-tac escapement has pallets embracing 2½ teeth and the pendulum has silk suspension. Circa 1820. Height: 16'' (40.5 cms.) Photo courtesy C.L. Papworth.

Fig. 55. French great wheel skeleton clock with a large going barrel mounted in the centre, sunburst pendulum and tic-tac escapement. Note the diamond pattern to the top of the frame, a characteristic feature of most of these clocks. Height: movement only, 11'' (27.8 cms.) Photo courtesy Sothebys, Billingshurst.

Figs. 56a, b, c. French gilded Y frame skeleton timepiece of 14 days duration with tic-tac escapement, ten spoke great wheel, and that interesting feature Wolfs tooth gearing which may be seen in the close up of the centre wheel. (56c) Height: 14'' (35.5 cms.)

Fig. 57. A French inverted Y framed skeleton timepiece with large 'scape wheel, open spring barrel, delicately pierced out hands, sunburst pendulum with tic-tac escapement and rosettes covering the heads of the pillar screws.

Fig. 58. French skeleton clock with skeletonised barrel and fluted and gilded inverted Y frame. The 10 spoke great wheel, some 6¾'' in diameter, has Wolfs teeth. Interestingly the enamelled dial is signed P. Renou Fecit à Paris. Invt. However no such maker is recorded as working in Paris at that time although there were two in London and it may be that one of these had emigrated from France.

Figs. 59a, b. Great Wheel skeleton clock with silk suspension and tic-tac escapement. The train count is:

Great Wheel	372:10	
Centre Wheel	276:6	
'Scape Wheel	80:	

Height: 13'' (33 cms.)

Fig. 60. An 8 day French chain fuzee skeleton timepiece with pierced barrel cover and the 'scape wheel mounted to the top right of the frame. It is signed on the chapter ring by Jefferys and Gilbert and would date from the 1820s. Height: clock only—10½'' (27 cms.) Photo courtesy Strike One.

Fig. 61. A good quality Y framed chain fuzee timepiece with skeletonised barrel and gilded plaques let into the base depicting classical scenes. The dial is surrounded by laurel leaves and serpents and above it and to the left may be seen the pin wheel of the pendulum controlled escapement. Photo courtesy Keith Banham.

Fig. 62. A flat Y framed skeleton timepiece formerly in the collection of Major Heathcote with decorative hooped spokes to all the wheelwork. The layout is particularly attractive with a large seconds beating balance wheel with fast/slow regulation to the top left and the 'scape wheel situated just inside the inner aspect of the chapter ring. The centre sweep seconds hand shows dead seconds. Height: 14¾'' (37.5 cms.) Photo Courtesy Keith Banham.

Fig. 63a.

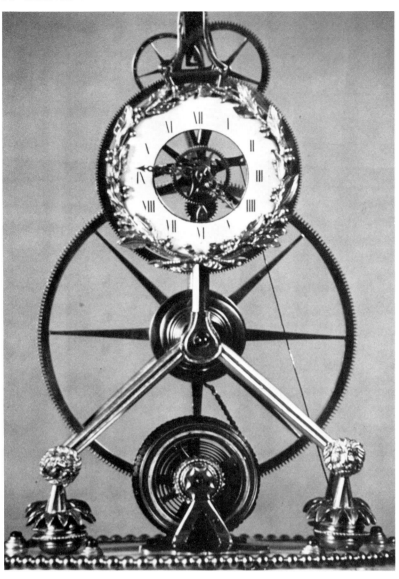

Figs. 63a, b, c. A fine French skeleton clock with a 5¼''
eight spoke great wheel, chain fusee and pin wheel
escapement, which is similar to but more complex than
that seen in Fig. 64. It has a 3-train movement with the
striking and chiming trains concealed in the base. The
quarters are struck on 8 bells and the hours on a separate
bell. Fusees are used for the strike and chime trains which
are controlled by count wheels. Thus the pull cord that
may be seen in the photo is not for repeat but to
resynchronise the quarters and hours should this become
necessary.

The visible time train is linked to the other two trains in
the base by a thin chain connected to a brass arm behind
the dial which is lifted at the quarters by 4 pins on the
minute wheel. An indication of the fine quality of this
clock is that the chain is helped to run smoothly by tiny
pulleys that guide it along the frame, down the right front
leg and into the housing.

Fig. 63b.

Fig. 63c.

Fig. 64. A small number of these beautifully conceived and executed skeleton clocks were produced in France in the late 18th/early 19th century. All appear to have employed a fusee in conjuction with a skeletonised spring barrel and pin wheel escapement. Whereas some, such as that seen here, were simple timepieces, in other instances either a quarter chiming or an independent musical movement was contained in the base. Height: 15½'' (39 cms.) Photo courtesy Peter Ineichen, Zurich.

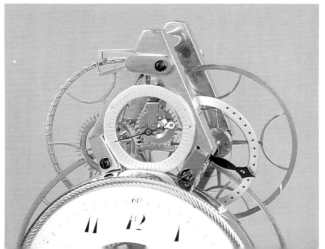

Figs. 65a, b. A skeleton clock similar in many ways to that shown in Fig. 62 but with the 'scape wheel (Fig. 65b) situated to the top left of the frame and the large seconds beating balance to the right. Instead of having a centre sweep seconds hand this is situated on a subsidiary dial immediately above 12 o'clock and just to the right of this a little of the fast/slow regulating lever may be seen.

A long lever runs from the centre of the balance wheel to the pallet arbor and either end of this lever is visible in the two close up photos. It is signed on the frame William Bert and dated 1826. Height: 13½'' (34.3 cms.)

GLASS PLATED SKELETON CLOCKS

This series of skeleton clocks bears a fairly close relationship to the Great Wheel clocks just described in that they also always have a large great wheel; they have the dial centre omitted and usually employ a pin wheel escapement. In the vast majority of cases they are spring driven but occasionally a weight driven example may be seen.

These clocks were probably all made by the same manufacturer but are almost always unsigned, although one has been seen bearing the name P.J. Beliard à Paris. They are frequently of long duration and sometimes have strike work; however in these instances, rather than provide a separate mainspring, power for the strike is often taken off the single going barrel via a remontoire which winds a subsidiary spring. This has the practical disadvantage that the strike cannot be corrected by more than a few hours because of the limited power of the small spring; the only way of overcoming this being to start the clock at the right time or to lock up the strike until it is synchronised with the time or only an hour or two behind, when it can be corrected. On occasions quarter striking glass plated skeleton clocks may be seen (Figs. 74 and 75) but they are comparatively rare as also is the incorporation of calender work. Interestingly on all the clocks which the author has seen which do incorporate this latter feature the revolutionary calender has always been employed which would indicate a date for these clocks in the early 19th century. The wheel work is of the highest order and is often amazingly thin, some being only 0.4 mm which makes one wonder how they survived so long. In nearly all instances Y shaped spokes, usually five in number, are employed. Like Great Wheel clocks Wolfs tooth gearing is sometimes used.

It is not uncommon to find that the glass plate has been replaced or that the bottom corners, where it is fixed to the base have been damaged and these have been made good with brass brackets. The Finish of the glass plates vary considerably; some are left plain but in other instances they are either painted round the edges, engraved or even have gilded brass fretwork applied to them.

The arbors for all the wheelwork run in either steel or more usually brass bushes let into the plates. One of the problems which may arise as a result of this is that, because the glass plate is relatively thick, the bushes and thus the bearing surfaces for the pivots are quite long and if any arbor is a little out of truth binding can occur.

Fig. 66. French glass plated skeleton clock with 5 Y shaped crossings; 10" great wheel and pin wheel escapement. Circa 1820. Height: 18" (46 cms.) Photo taken by courtesy of Sotheby's, Bond Street.

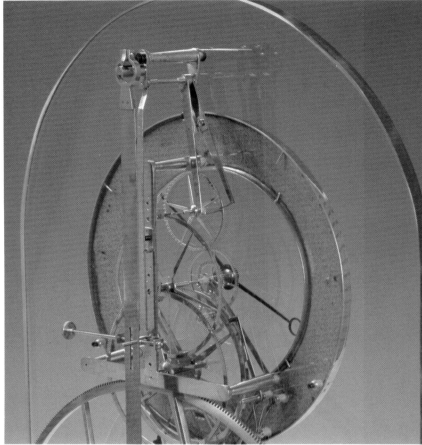

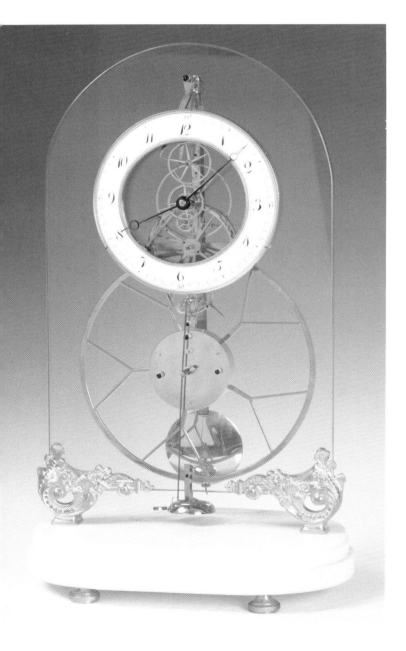

Figs. 67a, b. A fine unsigned glass plated skeleton clock with a single going barrel which is of six months duration. It has a pin wheel escapement and a steel rod pendulum with steel strip suspension. The remontoire train, which rewinds the small auxilliary spring for the strikework, is released every hour and a long linkage connects the countwheel striking mechanism to the bell hammer just above the base.

Some idea of the extreme delicacy of the wheelwork may be gained from the close up photograph which also shows the 'scape wheel. The thinnest wheel is only 0.4 mm thick. Height: 20½'' (52 cms.) Plate size 17'' x 10'' (43.2 x 25.4 cms.)

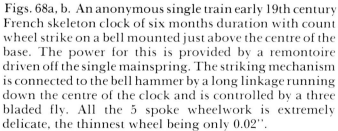

Figs. 68a, b. An anonymous single train early 19th century French skeleton clock of six months duration with count wheel strike on a bell mounted just above the centre of the base. The power for this is provided by a remontoire driven off the single mainspring. The striking mechanism is connected to the bell hammer by a long linkage running down the centre of the clock and is controlled by a three bladed fly. All the 5 spoke wheelwork is extremely delicate, the thinnest wheel being only 0.02''.

The delicate 'scape wheel has its pins facing to the rear and the steel rod pendulum rests on a knife edge. The glass plate has a gilt border and there is gilt and blue decoration over the mainspring barrel. The glass plate has been damaged at the bottom corners and this has been made good with brasswork. Glass plate 17½'' x 11'' x ⅜'' thick (44.5 x 28 x 0.95 cms.) Collection Albert Odmark, U.S.A.

Fig. 69. Glass plated skeleton clock of 14 days duration with pin-wheel escapement, steel rod pendulum with heavy brass bob and a large sun-burst rosette around the winding square. The brass framed glass shade protecting it may well be the original. Height: 17'' (43 cms.) Photo courtesy Ineichen, Zurich.

Fig. 70. A simple long duration glass plated skeleton timepiece with attractively crossed out wheelwork, an unusual style of hands and a fully fretted out applied gilt chapter ring. It rests on a gilt metal base with applied fretwork and is seen here without pendulum. It employs anchor escapement with a large 'scape wheel and very delicate pallets. Circa 1820. Photo courtesy Jean-Claude Sabrier, Paris.

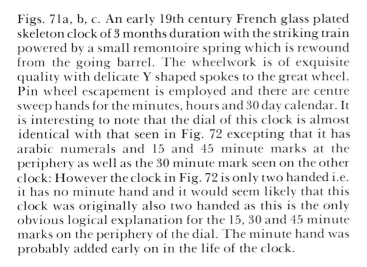

Figs. 71a, b, c. An early 19th century French glass plated skeleton clock of 3 months duration with the striking train powered by a small remontoire spring which is rewound from the going barrel. The wheelwork is of exquisite quality with delicate Y shaped spokes to the great wheel. Pin wheel escapement is employed and there are centre sweep hands for the minutes, hours and 30 day calendar. It is interesting to note that the dial of this clock is almost identical with that seen in Fig. 72 excepting that it has arabic numerals and 15 and 45 minute marks at the periphery as well as the 30 minute mark seen on the other clock: However the clock in Fig. 72 is only two handed i.e. it has no minute hand and it would seem likely that this clock was originally also two handed as this is the only obvious logical explanation for the 15, 30 and 45 minute marks on the periphery of the dial. The minute hand was probably added early on in the life of the clock.

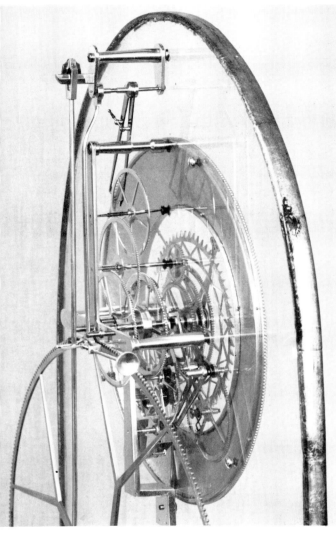

Fig. 71c.

Fig. 72. A fine quality single train French skeleton clock of long duration with single arm pin wheel escapement. The lancet shaped glass plate is very attractively decorated with delicate gilt fretwork which is also applied to the base. The steel rod pendulum has a knife edge suspension.

An unusual feature of the clock is that it has no minute hand but incorporates a centre sweep date hand for which the chapter ring is marked out accordingly. The wheels at the top of the train are only 0.018'' thick. Height: 19'' (48.2 cms.)

Fig. 73. The glass plate of this weight-driven clock is attractively engraved around the edge and as is usual with this type of clock has brass inserts which provide the pivot holes for the front of the arbors with the rear ends resting in a vertical brass rod which goes up the back of the clock. The 'scape wheel employs 72 teeth and has 6 crossings and its pinion meshes with a minute wheel of 276 teeth. The great wheel which is attached to the barrel has 5 Y-shaped spokes and revolves once every 26 hours and 40 minutes. Courtesy Musée des Techniques C.N.A.M., Paris.

Figs. 74a, b, c, & 75. The glass plated skeleton clocks seen in these illustrations have many similarities and would undoubtedly have been made by the same hand. Both have applied glass chapter rings with gilt numerals; centre sweep hands for minutes, hours and 30 day revolutionary calendar and a double ended hand for the days of the week, which goes round once a fortnight.

Their duration is 8 days and they have quarter striking on two bells and full hour strike on a single bell. Those seen in Fig. 74 are of conventional bell metal whereas the bells on the clock in Fig. 75 are made of glass and may just be seen behind the figures III and IX on the chapter ring.

The strike train gains its power from a small remontoire spring housed in a barrel at the centre of the great wheel. It is rewound from the going or time train barrel via the centre wheel pinion. To prevent the remontoire spring, which is being continually wound by the time train, being overwound, one end is left free thus allowing it to slip when it becomes fully wound.

The only problem with a small remontoire spring for the striking train is that it usually only has a sufficient reserve of power to strike 2-3 hours: Thus allowing it to strike each hour to preserve the time and strike sychronisation because the clock employs a count wheel, it will run out of power and cease to strike. If the hands are then advanced further the strike will go out and stay out even after the clock has been restarted and the remontoire spring has been recharged sufficiently for the clock to commence striking again.

To overcome this problem the clock seen in Fig. 74 is fitted with a ratchet which allows the remontoire spring to be rewound manually when necessary, a most interesting and useful feature which has not been seen in any other clock.

The primary differences between the two clocks are that the one seen in Fig. 75 employs a nine rod gridiron pendulum in conjunction with dead beat anchor escapement and is contained in a "four glass" case with levelling screws whereas that shown in Fig. 74 has a balance wheel beating 90 to the minute in conjunction with a flat hairspring and dead beat anchor escapement. This clock rests on an attractive oval gilded base overlaid with fretwork. Fig. 74 Private collection. Fig. 75 Courtesy Musèe des Techniques C.N.A.M., Paris.

KEYHOLE FRAMED SKELETON CLOCKS.

This series of clocks, which are almost invariably unsigned, were made in France in the first 20-30 years of the 19th century and are probably the most completely skeletonised ever made except possibly the glass plated clocks which have very similar movements and probably come from the same stable.

The chapter rings, usually with Arabic numerals, are very narrow, the hands delicate and the top part of the frame follows the contour of the chapter ring and lies behind it, thus making it virtually invisible from the front.

A narrow brass bar runs up vertically between 6 and 12 o'clock to carry the delicately executed train of wheels and extends downwards to support the spring barrel and, when it is provided, the hammer and bell.

Whereas most of the glass plated clocks are spring driven, with the keyhole clocks an appreciable number employ two weights the lines from which both feed onto a common central barrel.

As with the glass plated clocks Wolfs tooth wheelwork is sometimes used, a pin wheel is generally the escapement of choice and when strikework is provided power for this is supplied by a small auxilliary spring which is rewound by a remontoire driven off the single going barrel.

Examples of keyhole framed skeleton clocks may be seen in the Ecole d'Horlogerie, Dreux and the Conservatoire des Arts et Metiers, Paris.

Figs. 76a, b. An unsigned early 19th century French skeleton timepiece with the lines from the two driving weights feeding onto a central barrel. The five wheel train has a large great wheel and a pin wheel escapement is employed. The enamelled dial some 7'' in diameter has inner and outer engine turned gilded brass bezels. Height: 16'' (40.5 cms.) A similar clock but employing anchor escapement was sold by Herve Cheyette in the auction in Paris in June 1986. Photo courtesy Sotheby's, Bond Street.

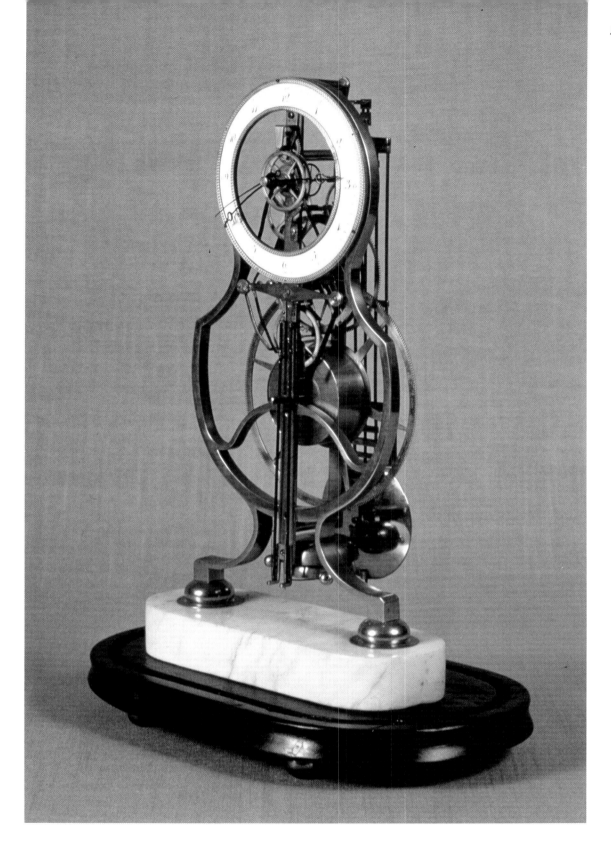

Fig. 77. Although the shape of the frame of this clock is different from the others in this small series which we have loosely termed "keyhole Skeleton clocks" the overall concept is very similar, for instance, to that seen in Fig. 78 and also some of the glass plated clocks. It is of 3 months duration and employs a remontoire for the striking train which has ting-tang quarter and hour strike on two bells mounted horizontally at the base of the movement. A pinwheel escapement is employed in conjunction with spring suspension to the half seconds beating 5 rod gridiron pendulum and thus the centre sweeps second hand revolves twice a minute. Height: 15¾". Private collection.

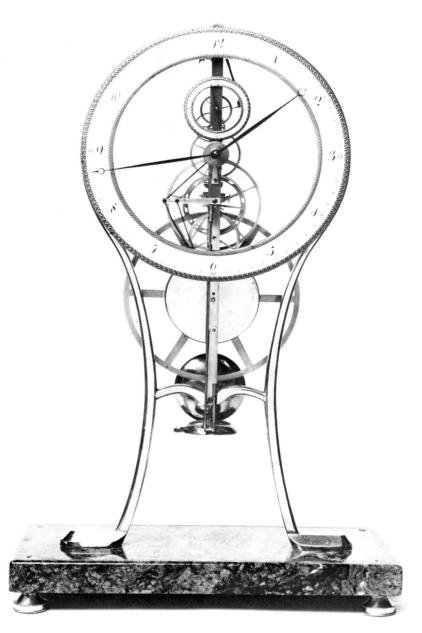

Fig. 78. A month duration keyhole framed skeleton clock formerly in the collection of Major Heathcote. It employs a pin wheel escapement which may be seen in the centre of the very narrow seconds ring, delicate Wolfs tooth wheelwork and lantern pinions. The strikework, the bell for which may be seen mounted horizontally just below the pendulum bob, is powered by a small auxilliary spring which is rewound by a remontoire from the single large going barrel. Height: 6'' (40.5 cms.)

Fig. 79. A more complex keyhole framed skeleton clock than the two already described. It has a 9 rod gridiron pendulum, pinwheel escapement and two driving weights which supply power directly to the going train and indirectly to the striking train via a remontoire which winds a small auxilliary spring. Like the others it is unsigned, but interestingly it is known to have been made in 1817 and was presented to the Musee du Conservatoire National Des Arts et Metiers, Paris (in whose collection it still resides) by J. Andreoud in 1885. Photo courtesy Musée des Techniques C.N.A.M., Paris.

Fig. 80. Although this anonymous weight driven clock which is in the possession of the Musee D'Histoire Des Sciences, Geneva, who kindly supplied the photograph, does not have a keyhole shaped frame it is very similar in most other respects to that shown in Fig. 76a and for that reason is included here. Minor differences are the size of the great wheel and chapter ring and the heavier driving weights.

PORTICO OR 4 PILLAR CLOCKS

This series of skeleton clocks, built in appreciable numbers in the period 1830-1850, are similar in their layout to a turret clock in that the two trains are laid out horizontally. They always employ a mock gridiron pendulum and usually have the bell mounted above the movement, although occasionally it is to be seen below and between the two spring barrels. The detail finish of the case varies, some for instance having wooden pillars and some gilt metal and the dial centre is often omitted.

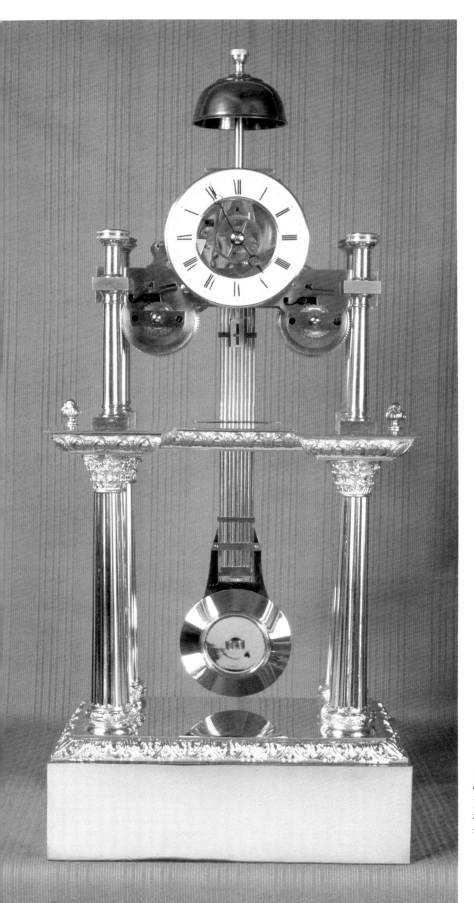

Fig. 81. French two train pillar clock of three weeks duration with Brocot dead beat escapement with steel pallets and a nine rod mock gridiron pendulum. It has an articulated linkage to the hammer which strikes on a silvered bell. Height: 22" (56 cms.)

Fig. 82. (Left) A similar clock to that seen in Fig. 81 but with a solid dial centre and the bell mounted below the dial.

Fig. 83. (Right) A two train French skeleton clock with plates fretted out in the form of the entrance gates to a church or a cathedral which would have been particularly apt as it was presented to the Reverend Fredk. Quarrington M.A. by his parishioners in 1851.

This clock would seem to be as close as the French got to using buildings as the basis for the design of their clock frames. However they did on occasions depict various buildings, mainly cathedrals, "in the solid" such as that seen in Fig. 84.

ARCHITECTURAL CLOCKS

In the second half of the 19th century the French produced a series of clocks based on some of their more important buildings, usually cathedrals, much as York Minster, Lichfield and Westminster cathedrals were being copied in England.

Fig. 84. A French "architectural" clock in which a cathedral is depicted in very fine detail in the solid. Photo courtesy Christies, King Street.

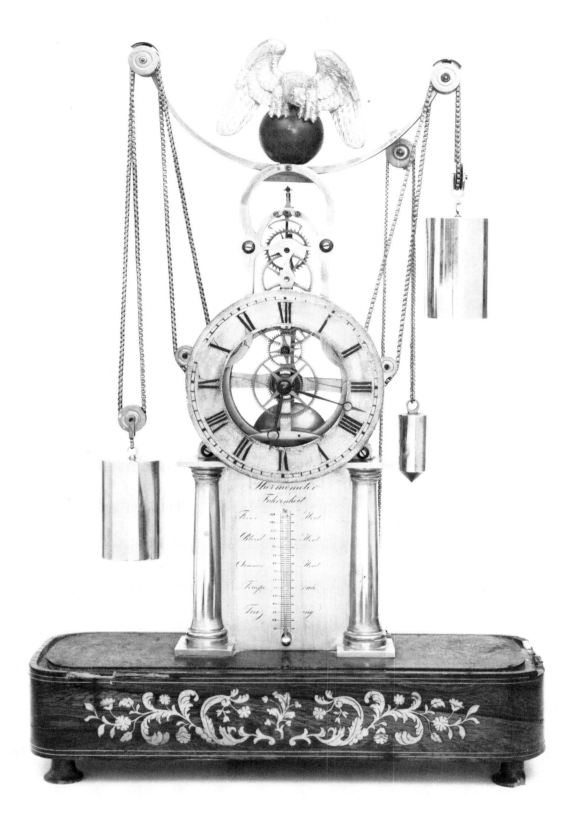

WEIGHT DRIVEN "EAGLE" CLOCKS.

Fig. 85. This group of clocks probably produced in the 2nd quarter of the 19th century, appear to be somewhat more complex than they actually are. In effect they are simple timepieces with anchor escapement driven by a single weight, that seen to the right of the clock. The one on the left has been put there for purely aesthetic reasons and the smaller one is a counterweight. Set beneath the silvered brass dial is the thermometer.

Exactly how many of these clocks were produced is not known, although the author has seen an appreciable number over the last 10-12 years. The majority rest on an inlaid rosewood base similar to that seen here. Somewhat confusingly some have appeared in the States bearing Willards name, although how this has come about is difficult to say. It is possible he imported them into the States and marketed them under his own name but this seems a little unlikely. Photo courtesy Sothebys, Bond Street.

YEAR DURATION CLOCKS.

In the second half of the 19th century great interest was shown in the production of clocks of year duration. One of the most interesting of these was that designed by Honere Pons of Paris in 1857 (Fig. 86). It employed a compound pendulum with point suspension and a single going barrel. Other manufacturers employed a different approach, using anything from two to nine mainsprings to provide sufficient power to keep the clock going for a year on one wind. In the case of the clock seen in Fig. 88, 18 mainsprings are used in all, 9 for each of the going and striking trains.

A somewhat earlier year clock has already been shown in Fig. 13, but probably the best known examples are the "400 day" clocks with torsion spring suspension which have been produced in many hundreds of thousands if not millions from the late 19th century to the present day. Their secret of minimal power absorption lies in their very slow "beat", a torsion pendulum approximately 6" long making only one excursion every four seconds.

Figs. 86a, b. *Year Clock by Honere Pons of Paris.*

Some idea of the efficiency of Pons clock may be gained from the relatively small mainspring which he employed. The most important factor in his design is probably the use of an equipoise seconds beating pendulum which is supported by two steel points which rest on hardened steel cups. Unfortunately although this pendulum absorbs very little power it depends on constant friction if good timekeeping is to be obtained and this is almost impossible to achieve, particularly on a relatively simple clock like this with a going barrel.

The escapement is an unusual form of deadbeat with long pointed pallets coming out from the anchor almost at right angles. Immediately below the main dial is an equation of time table which seems more than a little optimistic for what must surely be a relatively inaccurate, although fascinating clock. For a much fuller account of it the reader is referred to Richard Goodes article which appeared in the Horological Journal.

Figs. 87a, b. A French year duration clock with perpetual calendar, moon phase and equation of time dials to the bottom left and aneroid barometer to the bottom right. It has 5 going barrels and Brocot visible escapement. This clock, which was in the collection of Major Heathcote, was skeletonised by Peter Bonnert in 1973. Photo courtesy Christies, London.

Figs. 88a, b. A year duration skeleton clock with engraved front plate, Brocot visible escapement and no less than 18 spring barrels, 9 for both the going and the striking trains. Each row of barrels is wound from the front using a male key. The subsidiary dials show days of the week and month, months of the year with their zodiaccal signs, moon phase and equation of time. The backplate is signed by Brocot and numbered 5457 and the calendar dial is also signed but the main dial bears the name James Simpson, Secunderabad, the retailer. Height: 17'' (43 cms.)

Figs. 89a, b. French perpetual calender skeleton clock with count wheel strike, signed V. Leboef à Deville 1884. Photos courtesy Jean-Claude Sabrier, Paris.

THE 'GREAT EXHIBITION' AND OTHER SIMILAR CLOCKS

By the late 1830s the production of most of the skeleton clocks already described such as the "Great Wheel, Glass plated, Key hole framed and Eagle clocks had largely ceased and only a handful of makers were still producing a very limited number of fine quality skeleton clocks on a one off basis. However by the 1840s an attractive range of small skeleton clocks started to appear of which by far the best known are those seen in Figs. 90 and 93. The majority had an engraved frame and pendulum bob, an enamelled chapter ring with an alarm disc (Fig. 90) in the centre of this and the actual alarm bell built into the base. They were also made without an alarm (Fig. 93) and sometimes the frame was left plain. Attractive features were the engraved disc at the top of the frame for fast/slow regulation and the fact that the alarm could be wound and set by pulling two cords which came out below the base of the clock.

Victor-Athanase Pierret showed these clocks at the Great Exhibition in 1851 and his name frequently appears on their frames. It has been stated that he sold some 10,000 at the Great Exhibition but this would not appear to be correct. In the publication Horlogerie—"Outillage et Mechanique" de V—A Pierret published in Paris in 1891 by the Imprimerie Nouvelle (Association Ouvriere) pages 12-15 it is stated that Pierret did indeed first exhibit the clocks in London in 1851: However it goes on to say that 12,000

Fig. 90. A miniature French skeleton clock of the type commonly known as "The Great Exhibition Skeleton Clock" because Pierret exhibited them for the first time at the Great Exhibition in London in 1851 where they proved to be very popular.

It has an engraved frame and pendulum bob and an enamelled chapter ring with the engraved and silvered alarm disc visible within it and the fast/slow regulation for the silk suspension above. The cord on the right of the base is used to wind the alarm and that on the left is for setting it by means of a pawl which can be seen inside the chapter ring between VIII and IX o'clock, thus obviating the need to remove the dome. The bell for the alarm is incorporated in the base and may just be seen extending below it. It bears the stamp on the base Ms.Honorables Exp. Paris, Londres. Height: 10" (25.3 cm).

were produced since that time i.e. between 1851-1891 and that of this 12,000 some 10,000 were sold to English clock dealers. He concludes by expressing the hope that they will continue production for a lot longer, implying that they were still making a considerable number of the clocks at this time (1891). Shown in Figs. 91 and 92 are the drawings of the skeleton clock included in Pierrets 1891 publication.

Pierret was born in 1806 at Bucy-les-Pierrepont. At 13 he was apprenticed to Rollin and at 24 he established himself at the Rue des Bons-Enfants. He made calender clocks, became interested in compensated balances for chronometers and besides the skeleton alarm showed a planetarium clock at the Great Exhibition. In 1865 he gave up his business in Paris and retired to Neuilly where he continued his research until he died in 1893.

Rollin, who worked in Paris from 1840-70, first at the Rue de la Corderie (1840-50) and later at the Rue de Bretagne (1860-70) also produced a small timepiece alarm skeleton clock in appreciable numbers (Fig. 94). The overall concept was very similar to Pierrets but the frame, which was not so decorative, was left plain and the alarm bell displayed at the base of it.

Small two train skeletons clocks were also being made at this time of which the most attractive is probably that seen in Fig. 95 but in these instances only a single mainspring was usually employed to drive both trains. Calenderwork was sometimes incorporated in the design (Fig. 96) and on occasions they were mounted on ornamental gilded columns (Fig. 97).

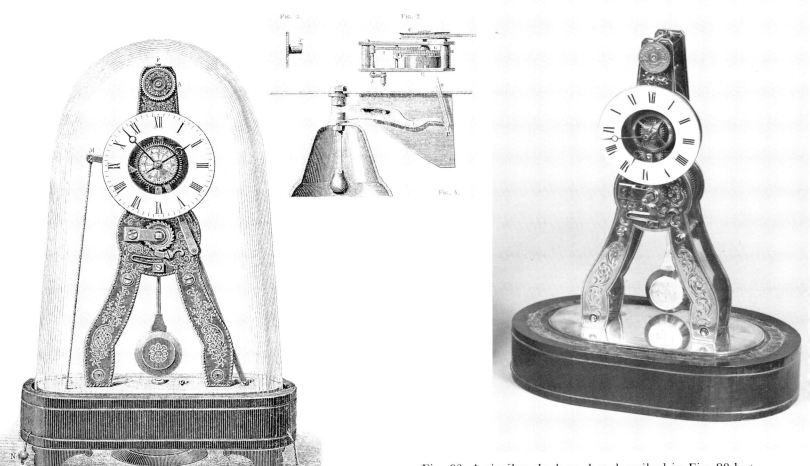

FIG. 1. — Pendule squelette à échappement, à repos et à réveil.

Figs. 91 & 92. The drawings which appeared in "Outillage et Mechanique de V—A Pierret" 1891 showing the "Great Exhibition" skeleton clock and its' ingenious alarm.

Fig. 93. A similar clock to that described in Fig. 90 but without alarm. Because of the omission of the bell from the base feet are not required and thus the clock is a little smaller. It has a brass strung rosewood base and is stamped on the brass base plate Ms.Honorables Exp. Paris, Londres and the back of the frames bears the initial P and the number 3159.

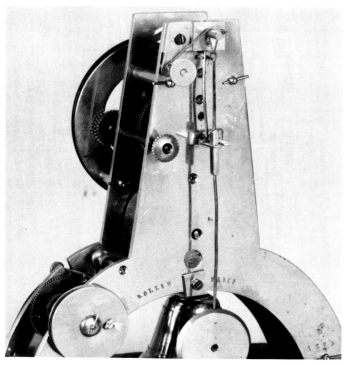

Figs. 94a, b. A timepiece/alarm skeleton clock with inverted Y frame having the bell visible immediately beneath it. The spring barrel for the alarm may be seen to the bottom right of the frame and this is wound by a cord passing out through the base. However unlike the clock shown in Fig. 90 the dome has to be removed to adjust the silk suspension or set the alarm. It is signed on the enamelled chapter ring which has blue numerals, Rollin à Paris and this signature is repeated on the backplate where it is also numbered R 1283. The plates bear the stamp of Japy Freres who presumably made the blancs Roulants.

Fig. 95. An interesting mid-19th century French skeleton clock standing on turned brass feet and with a well executed simulated marble base. It has a single spring barrel with two greatwheels attached to it, one of which drives the going and the other the striking train the countwheel for which may be seen mounted on the frontplate. Height: 12" (30.5 cms.)

Fig. 96. This anonymous clock is similar to that seen in Fig. 95 in that it has a single spring barrel for both the time or going and the striking trains. The former is driven by a gear on the front end of the barrel and the latter by a wheel on the back of it. The frontplate, base plate and barrel cover are all decorated with foliate engraving. The dial to the bottom left is for the alarm, the winding arbor for which is situated below and to the right of it. Different hammers are used for the strike and alarm but there is only one bell which is mounted in the base. The dial on the right indicates the days of the month. Silk suspension is used and a Brocot escapement with steel pallets. Collection A. Odmark, U.S.A.

Fig. 97. A similar skeleton clock to that just described in Fig. 96 but with the movement supported by four gilded columns within which is a gilt bronze cherub holding the inverted bell for the strike and alarm. Photo courtesy Jean-Claude Sabrier, Paris.

Figs. 98a, b. An interesting month duration French skeleton timepiece with a substantial inverted Y Frame, pinwheel escapement, beat regulation and a steel rod pendulum with heavy brass cased lead bob. It is signed on the enamelled chapter ring Jouflinot R. Du Temple 25 (Paris). The mainspring is also signed and dated 1846. Tardy records a Jouffroit working in Rue Vieille du Temple in 1840 and one cannot help wondering whether this was the same man. Height: clock only—10'' (25.3 cms.) Overall Height: 15'' (38.1 cms.)

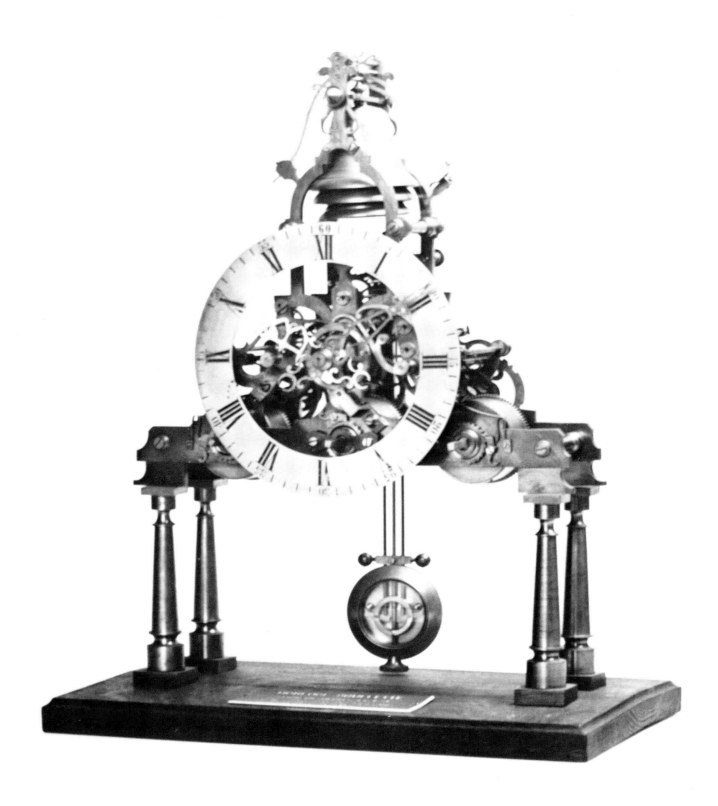

Fig. 99. A three train skeleton clock chiming the quarters and striking the hours on 3 bells. The spring barrels are mounted horizontally rather like a turret clock; it has a Brocot escapement and a half seconds pendulum with Ellicotts system of compensation. Made by A. Beillard and his pupils circa 1900. Height: 22½'' (57 cms.) Photo courtesy Jean-Claude Sabrier, Paris.

UNUSUAL CLOCKS

The French undoubtedly used more imagination and ingenuity in the design and execution of their clocks and the materials they employed than any other nation and the clocks illustrated here have been chosen to try and give some idea of the diversity rather than the complexity of their designs.

Figs. 100a, b. *A Skeleton Clock with Glass Bells.*

A mid-19th century three train skeleton clock with the barrels laid out horizontally as on a turret clock. The movement has a pin wheel escapement and a half seconds beating pendulum with Ellicotts system of compensation clearly displayed in the centre of the glazed bob. The main dial which is signed Detouche à Paris shows the minutes and hours and the subsidiary dial below it seconds. Detouche, a fine and ingenious maker, worked on electro-magnetic pendulums as early as 1851 and produced many interesting devices in association with such makers as Robert Houdin, Oudin and Brisbart Gobert. It has grand sonnerie striking using two bells for the quarters which are controlled by a snail and one for the hour which employs a count wheel.

A fascinating feature is that the bells, supported in a similar manner to the shades used on a Chandelier, are made of glass, a choice of material which initially seems foolhardy in the extreme until one remembers that cast bell metal is also extremely brittle. It has always seemed to the author a contradiction in terms and yet an undoubted fact that a bell can be hit hard by an iron hammer many millions of times during its' lifetime without breaking but that if you drop it only a fairly small distance it will invariably crack or disintegrate. It is, of course, the extreme hardness, even brittleness of a bell which gives it its' tone and this must be equally true whether it is made of metal or glass. Collection Dr. S. P. Lehv, New York.

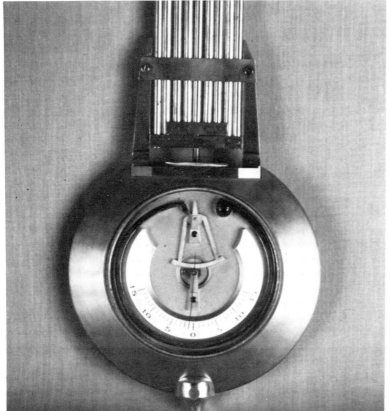

Figs. 101a, b. A late 19th century French skeletonized table regulator with gilded plates and wheelwork; six silvered brass pillars, which seem to be a hallmark of this makers work, and Brocot escapement. It has a silvered brass chapter ring some 7½'' in diameter with Roman numerals. There are centre sweep minute and hour hands; seconds are indicated below 12 o'clock and there are subsidiary dials at 9 o'clock and 3 o'clock for day and date respectively.

The nine rod gridiron pendulum is beautifully displayed and executed with a bimetallic strip running around the inner edge of the bob. This presses on a cam connected to a rack which rotates the pinion to which the temperature indication hand is fixed. There is a beat scale immediately below the bob. The clock is signed Ed. Francois. Her 14 Bd. des Filles du Calvaire (Paris) and was formerly in the collection of Gabriel Moreau who was director of the Ecole D'Horologerie D'Anet where the collection was housed.

Francois appears to have been both a fine and ingenious maker, for instance a carriage clock by him having been seen which was of the highest quality, incorporating both a Chronometer escapement and a remontoire. Height: including base—15½'' (39.5 cms.)

Papier Mache Skeleton Clocks

Figs. 102a, b. Papier mache skeleton clocks are by their very nature fragile. Only a small number were made in the first instance and of these very few have survived. To find one such as that illustrated which is still in working order is truly remarkable as almost invariably the minimum deterioration found is loss of teeth from disintegration of the wheelwork.

The frame of this clock is painted white with gilt decoration and it is powered by a vertical coil spring, giving a duration of 30 hours, which is rewound through the base, thus obviating the need to remove the dome. Lantern pinions are employed the "wires" for which are turned of a dense wood for the two high speed arbors and of a white material which appears to be bone for the lowest speed (the centre wheel). The supports which provide the pivot holes are of horn and a similar material is used for the pallets which are dead beat.

The wood rod pendulum is supported by two pins in a similar manner to a knife edge and regulation is provided by a wooden weight which slides up or down the rod; this being marked "Advance at the top and Retard at the bottom".

The dial face of this clock is of paper and bears the inscription from XI to I "Cartorologe Invariable" which possibly implies that it has the virtue of being immune to temperature changes due to its construction in papier mache.

The bottom half of the dial is marked in gilt and bears the caption "a Paris-Chez Ch. Rouy, Brevete du Roi, Galerie Vivienne a l'uranorama."

A very similar clock bearing the same incriptions was advertised by Charles Allix in the Sept. '73 issue of Antiquarian Horology and his comment that "a l'uranorama" refers to the 15 year brevet obtained by Charles Rouy in 1816 for "une machine dite urano-graphique" and has nothing to do with the clock in question is of interest.

Two more papier mache clocks are illustrated by Royer Collard,[1] one of which is very similar to that shown here. Both bear the inscription N.S. Villemsen et Cie, Galerie Vivienne No. 38, Novelles Pendules en Carton Per-fectionees. Tardy in his Dictionnaire des Horlogers Francais records Villemsen et Cie as working at this address around 1830.

As Charles Rouy and Villemsen et Cie both worked at Galerie Vivienne and were producing similar clocks it seems likely that they either worked together or that one succeeded the other in business. If the latter is the case then it is probable that Villemsen et Cie succeeded Charles Rouy. (i) Royer-Collard B. Skeleton Clocks. Pages 133-135. N.A.G. Press 1969.

[1] Royer-Collard B., *Skeleton Clocks*, pages 133-135, N.A.G. Press, 1969.

Skeleton Clocks Made Of Ivory or Bone

Many of the decorative and unusual pieces made of ivory or bone produced in the 19th century originated in France whilst others are reputed to have been made by French soldiers imprisoned in England during the Napoleonic Wars: However so far as the author is aware there is no documentation of a clock having been produced by them. Nevertheless support is given to the belief that most skeleton clocks made of ivory or bone were produced in France by the appearance in the first chapter of "Manual Du Tourneur" of a fully detailed description of the manufacture of a skeleton clock made of ivory. This description running to some 15 pages (243—258) is far too lengthy to reproduce here but the drawing which accompanies it is shown in Fig. 103.

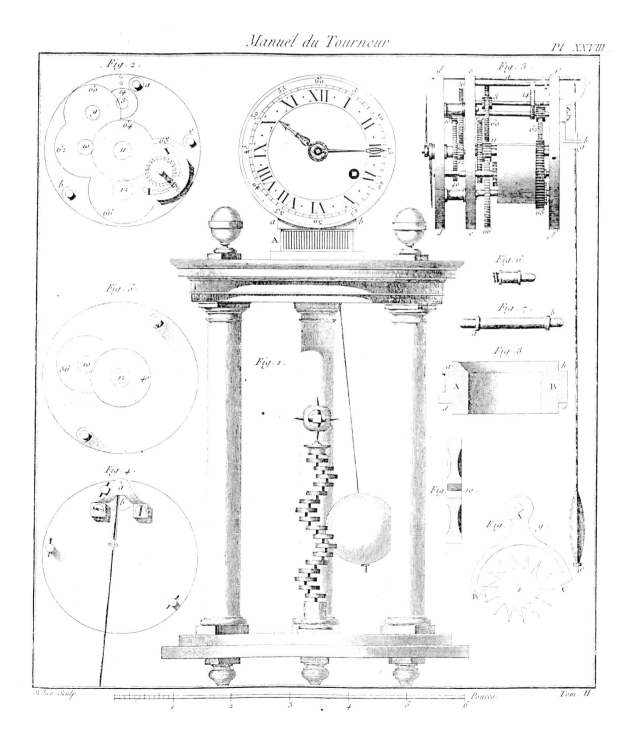

Fig. 103. An extract from Manuel du Tourneur PLXXVIII figs. 1-10 detailing the construction of an ivory skeleton clock.

Fig. 104. A rare little mid-19th century, single train fuzee skeleton clock made almost entirely of bone, including the pillars and pinions. It is shown prior to restoration in Pl 43 of British Skeleton Clocks.

The use of a fuzee does at first suggest an English origin: However the French did sometimes use fuzees on their skeleton clocks, such as those with a large Great Wheel.

Fig. 105. The delightful little clock seen here was featured in the Derek Roberts Antiques exhibition of skeleton clocks, exhibit No 42. Although it was in remarkably original condition no attempt was made to get it back into working order as this would have necessitated recutting or replacing all the wheelwork, and thus destroying its originality. The reason for this is that ivory and bone, like wood, shrink across the grain which results in all the wheels becoming slightly oval.

The clock is made in the shape of an elbow chair with the spring barrel in the base and the lovely little chapter ring a little above and in front of it. The wheelwork, all with 5 crossings, runs up the chair back with the 'scapewheel, also of ivory, at the top and even the pendulum is made entirely of ivory. The clock rests on a green velvet base which bears a plaque inscribed John Haynes. Cork. Height: clock only—4½'' (11.4 cm).

Figs. 106a, b. The attractive little fuzee timepiece shown here, which is only 10'' (25.4 cms.) high overall, has a seconds hand above the main dial which goes round every twelve seconds. Virtually every component except the hands, arbors and clickwork are made of ivory. It is signed above and between the two front columns "Edward Bates, Kingsland" and dated 1858. Dr. S.P. Lehv Collection, U.S.A.

Perpetual Motion

Man's dream of obtaining "perpetual motion" probably goes back thousands of years and has occupied the minds of many clockmakers as well as those working in other fields of human endeavour. Probably so far as horology is concerned the closest anyone has come to achieving this aim is James Cox who in the 1760's devised a clock which made use of changes in temperature and barometric pressure. This was achieved by the upwards or downwards movement of a column of mercury which, by means of a series of linkages, wound up a wheel. A clock based on similar principles is the modern atmos clock.

Many mechanical clocks have been devised which at first sight should achieve perpetual motion but all sadly are based on mathematical or physical inaccuracies. Probably the most convincing is that devised by Jean Geiser and his son David who worked in Neuchatel (Fig. 107). At least two similar clocks were produced, probably at a later date, which had carrying handles on top (Fig. 108a, b), possibly to simplify their transport from one Fair or Exhibition to another where they would be displayed as mechanical miracles.

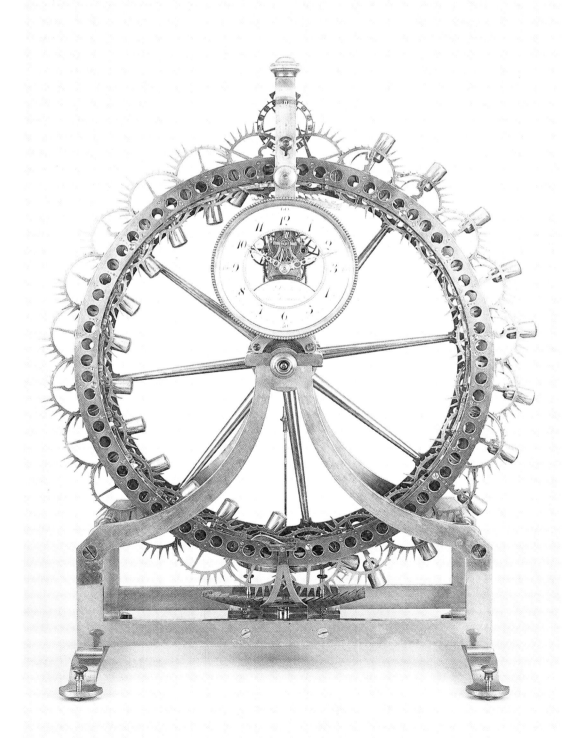

Fig. 107. For a description of this clock and an amusing account of one clockmaker's encounter with it we can do no better than reproduce an extract from *La Pendules Neuchâteloise* by Alfred Chapuis which gives a translation of an article which appeared in a German Scientific review in 1819 and the clockmaker's story by A. Mongrel in which the secret of the clock's perpetual motion is revealed. See Appendix, page 242. Photo and description courtesy Antiquorum, Geneva.

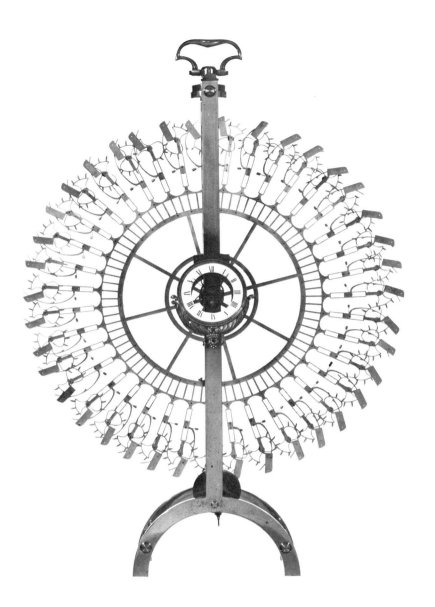

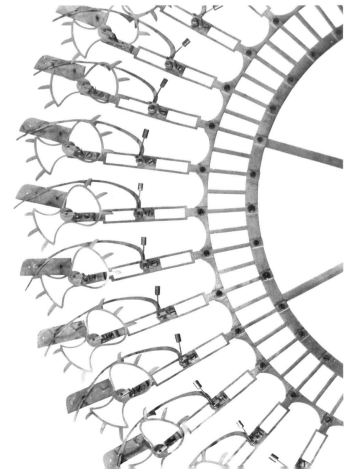

Figs. 108a, b. An anonymous "perpetual motion" clock which, judging by the carrying handle, was designed to be transported to fairs and exhibitions for public display. The mainspring is concealed in the central arbor and to rewind it the wheel has to be turned backwards some 24 times. It should be noted that as photographed the clock is set up wrongly in that the pivotted weights should only be outside on one side of the wheel. Height: 35½'' (90 cms.) Wheel diameter: 23⅝'' (60 cms.) Photos by Courtesy of Bernard Oger & Etienne Dumont, Paris.

Fig. 109. A most ingenious skeleton clock made by Brosse à Bordeaux. The driving force is provided by the outer steel crossbow which is drawn into tension by two chains acting upon a drum fixed to the great wheel above.

The enamelled dial rotates against a fixed central hand and is calibrated in hours, quarters and 5 minute divisions. The periphery of the small wheel above it is marked out in minutes and above that again may be seen the scape wheel.

A similar but more complex clock based on the same principle may be seen in Fig. 225 of the chapter on modern skeleton clocks. Photo courtesy Musée des Techniques C.N.A.M., Paris.

Fig. 110. One of the most ingenious skeleton clocks the author has ever seen. It has a caroussel with torsion suspension and anchor escapement and, quite fascinating, square wheels to the train and an hour wheel which is twelve sided, almost certainly a unique feature. The thought of cutting such wheels would fill most clockmakers with horror. I will leave it to the reader to work out exactly why such an amazing concept actually works! Photo courtesy Musée des Techniques C.N.A.M., Paris.

SKELETONISED CLOCKS AND ORRERYS

The margin between skeletonised and skeleton clocks is a narrow one and indeed some of those already described in this chapter may well be considered by some to be wrongly attributed, although this is probably of no great significance. However there are also clocks of relatively conventional design which have been deliberately skeletonised to display the clockmakers art to the full and a few of these are shown here. Also illustrated are some clockwork orrerys as by their very nature with these pieces everything is displayed as completely as possible and that seen in Fig. 116 could well classify as one of the most remarkable skeleton clocks ever made.

Figs. 111a, b. This attractive little 18th century pendule de voyage by Dubos, Paris is at first sight relatively conventional in design but on opening the back door it can be seen that the entire movement has been skeletonised as completely and meticulously as possible. It has strike, quarter repeat, alarm and calendar with a fuzee for the going train and Duplex escapement. Photos courtesy Christies, King Street.

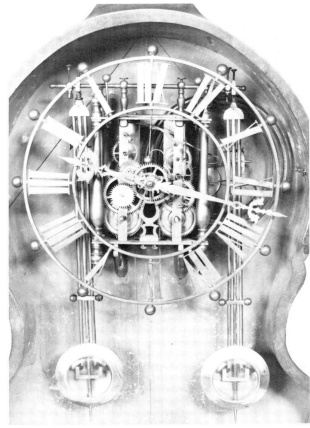

Figs. 112a, b. An interesting French provincial walnut mantel clock with original bracket, not seen here, into which the two weights descend. It has attractively turned brass posts; a fabricated brass chapter ring; recoil escapement and vertical rack strike on a gong.

A most unusual feature is that the two three rod gridiron pendulums are linked, presumably to try and even out any errors of compensation or regulation but somewhat surprisingly they have individual adjustment for beat.

Figs. 113a, b. The simplification and the reduction of the working parts of a clock to the minimum was one of the aims of many clockmakers in the 18th century. It was hoped that in this way friction would be decreased and efficiency and accuracy improved. One of the obvious ways of overcoming this problem was to reduce the number of wheels in the train, ideally to just one.

Pierre Le Roy is recognised as the inventor of the one wheel clock, although in fact the design consisted of two wheels, one slightly smaller than the other upon which it was superimposed. Lepaute also made "one wheel" clocks and examples were produced in Vienna.

The one shown here which is in the Ilbert Collection at the British Museum would date around 1775, and is contained in a panelled walnut case. It has a glass dial with gilt numerals which bears the inscription "Inventee et Executee par Le Roy Fils. Paris". The escapement is a form of dead beat duplex and the clock is driven by a weight with recessed pulley employing Huygens endless rope which gives a duration of 8 days.

The pendulum has double knife edge suspension and seconds are shown by a pointer which moves first to the left and then to the right of a cutout in a decorative sector situated below 6 o'clock. Photos taken by kind permission of the Trustees of the British Museum.

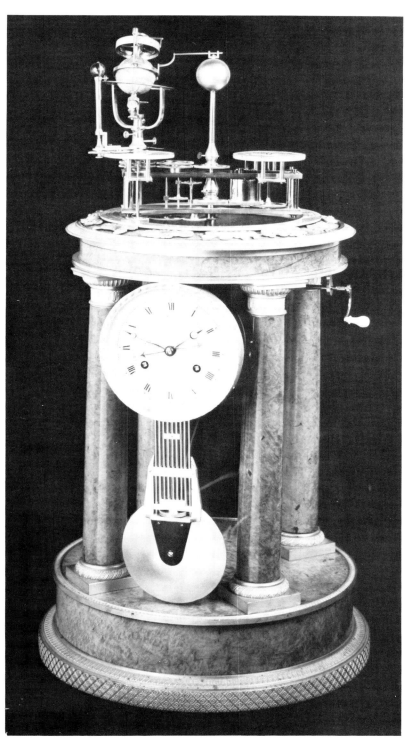

Fig. 114. An Orrery clock signed Leroy et Fils, Hrs du Roi, Raingo Freres Hrs. Bte du Roi à Paris. There is a silvered year calendar ring and gilt metal Zodiaccal symbols encircling the mechanical tellurium which is geared to the movement but may also be set by crank drive.

The tellurium revolves anticlockwise demonstrating the motions of the Earth and Moon in their rotation around the sun. It indicates the following:

1. The days of the month and the months of the year.
2. The position of the sun in the Ecliptic.
3. The Bissextile (leap year) cycle.
4. The age and phase of the moon.
5. The Siderial period of the moon.
6. The approximate declination of the sun.
7. The approximate times of sunrise and sunset in the Northern Hemisphere.
8. Solar Time.

Photo Courtesy Christies, King Street.

Fig. 115. A skeletonized Orrery clock of 8 days duration with going barrel, 4 wheel train, Sully type escapement with pin-wheel single locking roller and double sided L rings for seconds and meantime. The Orrery above, which it drives, has 5 pointed stars representing the planets and is enclosed in a glass globe etched with the constellations. Height: 20½'' (52 cms.) Photo Courtesy Christies.

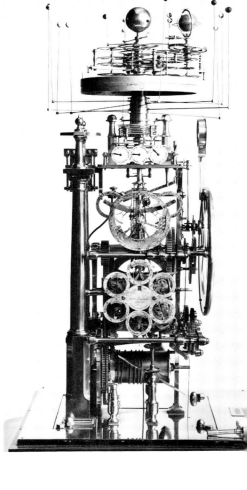

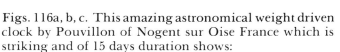

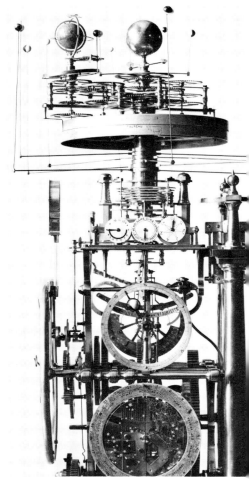

Figs. 116a, b, c. This amazing astronomical weight driven clock by Pouvillon of Nogent sur Oise France which is striking and of 15 days duration shows:

1. The revolution of all the planets around the sun and the revolution of the moon around the earth.
2. Mean time for the meridians of Greenwich and Paris.
3. Length of the day and night.
4. The time of sunrise and sunset.
5. The elevation of the sun.
6. True solar time and the equation of time.
7. The moons apogee and perigee.
8. The position of Ursa Major and Minor for an observer in the Northern Hemisphere.
9. The day of the week; the month; the seasons; leap years and Ecclesiastical Computer giving the: Epact, Dominical letter, Golden Number, Indication; Solar cycle; date for Easter and January 1st. Photos and details of this clock were kindly supplied by Jean-Pierre Rochefort, Paris.

Fig. 117. A large and very complex astronomical clock housed in a case which is fully glazed on all sides. There are 4 dials at the top painted on glass which indicate seconds, minutes and hours from each aspect of the clock and below this are all the astronomical indications which constantly give the relative positions of the sun, moon and earth. Most unusually the top dial gives the day of the week on which Feb. 29th falls for each of the leap years whilst that below and to the left of it indicates the 4 year cycle, being divided up into 48 months. Photo courtesy: Musée des Techniques C.N.A.M., Paris.

Fig. 118. This interesting clock was shown by its maker Aug Larible at the "Exposition Universelle in 1855. The escapement is a detached one which was invented by Larible and is illustrated and described in the catalogue Du Musée Des Arts Et Metiers Paris section J.B. pages 189-190.

The outer rotating ring comprises a year calendar with the zodiacal signs and may be read off against a fixed pointer at the bottom. The top inner dial gives the age and phases of the moon, that to the left shows the times of sunrise and sunset whilst the one on the right gives the equation of time (sun slower/sun faster). At the bottom mean time is indicated with the days of the week being given by a small rotating chapter above 6 o'clock. Bissecting this chapter ring and rotating against a pointer at the centre of the dial is shown universal time ie the time at various cities throughout the world. Photo courtesy Musée des Techniques C.C.A.M., Paris.

Figs. 119a, b, c, d, e. *Breguet Table Regulator.*

A skeltonised table regulator signed on the convex dial by Breguet, but unnumbered. It has two spring barrels which periodically lift the remontoire weight; thus in effect converting the clock from spring to weight driven.

It has marked similarities to the weight driven Lepine shown in Fig. 49a, b employing a similar gridiron pendulum, dial and hands. The dead beat escapement is also virtually identical but is mounted between the plates rather than behind the backplate. The frames and feet of the two clocks have only minor differences and the figures surmounting the dials are virtually identical.

REMONTOIRES

As early as the mid-16th century it was realised that if accurate timekeeping was to be achieved it was essential to provide the movement with a constant driving force. In the case of weight driven clocks this was no problem as the force exerted by a weight whether it is fully wound or nearing the end of its' run will be the same. However with a spring driven clock or watch the power will obviously decrease as the spring runs down.

The Fuzee And Stackfreed

The first two methods of overcoming this problem, which were both introduced in the first half of the 16th century, were:
a) *The Fuzee*, with which readers will be familiar as it became a standard feature of nearly all English and some continental spring clocks. The basic principle is the equalisation of the pull of the spring by using a tapering or cone shaped device onto which the line from the mainspring winds. In this way when the spring is fully wound it pulls on the narrow end of the fuzee and thus, because of the decreased leverage, its' force is reduced and as it gradually runs down and the line from the barrel moves along to a wider part of the fuzee its force is increased.
b) *The Stackfreed*. The principle of the stackfreed is basically the use of a checking spring which, by pressing on a disc of irregular outline exerts a variable pressure on the spring and thus equalises the force it applies to the train. The profile of this disc obviously has to match the characteristics of the mainspring being used and varied very appreciably from the time when it was first introduced in the 1520s to the time when it was going out of favour some 100 years later. Interestingly on some of the later stackfreeds the profile and set up was so adjusted that the driving force of the mainspring was actually increased as it neared the end of its' rundown.

Remontoires

The word remontoire is used in many different contexts, but in its' simplest form it means the rewinding of a driving weight or spring at set intervals. Thus the mainspring of a clock may be used to rewind a smaller spring every few minutes to keep it at the same state of wind and thus provide a relatively constant force on the movement. Alternatively the mainspring may be used to rewind a weight which then drives the train. This in many ways is the ideal as in effect it converts the clock from spring driven to weight driven which provides a constant driving force.

Many other forms of Remontoire are used, for instance although the striking train of a clock may be driven conventionally by either a mainspring or weight the going train is sometimes powered by a relatively small weight or spring which is automatically rewound every time the clock strikes and that train is released.

The reverse of the foregoing may be seen on the fascinating keyhole shaped skeleton clock in Figs. 123a, b, c. in which the mainspring powers the going train directly and rewinds a small auxilliary spring which provides the power for the striking train.

A relatively modern form of remontoire is the use of the electric motor which is triggered when the driving weight nears the end of its travel and automatically rewinds it.

Remontoire or Constant Force Escapements have the same aim as most of the remontoires already described in that they are devised to provide a constant force at the escapement and thus keep the balance wheel oscillating or the pendulum swinging at an even rate. A good example of this is Denisons Gravity Escapement in which arms are lifted up a constant distance on either side of the pendulum and then allowed to drop on and impulse it at a predetermined point in its' arc of swing; thus applying a constant force to it which is not influenced by the power coming through the train, which may vary.

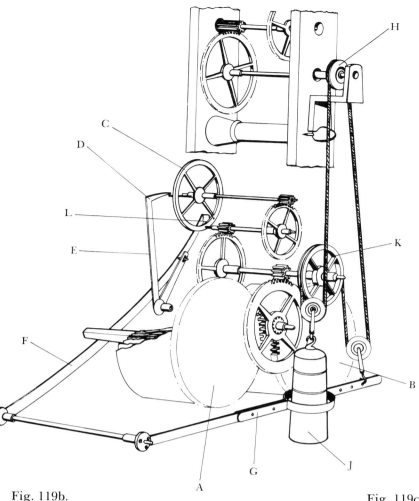

Fig. 119b.

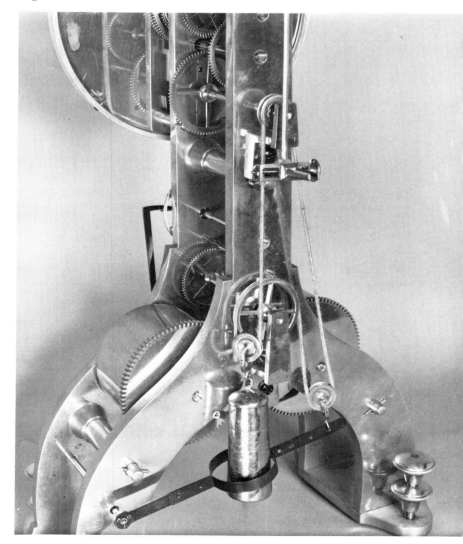

Fig. 119c.

The remontoire is driven from the twin spring barrels A and B which, through a conventional train of three wheels, drive a light flywheel C supported in a cock on the front plate. This flywheel carries a cross lever D.

Mounted on a post lower down on the front plate is the Vee shaped lever E which, on its longer arm, has a detent which locks the train by holding the lever D. The shorter arm of E has a slot through which runs a pin mounted on the long curved piece F. This piece is rigidly connected through the plates to another long piece G behind the backplate.

The escapement train of the clock is driven via an endless cord running over a small pulley H behind the backplate and actuated by the falling weight J. The cord in turn runs over a large pulley K which is firmly fitted to the outer end of the second arbor of the remontoire train, and held in a cock on the backplate.

With the remontoire train locked, the falling weight lifts the piece G which, in turn, slowly withdraws the locking detent on E and eventually releases the train. The flywheel C makes one full turn and in so doing the rotation of the pulley K lifts the weight a short distance and this lowers the piece G and causes the detent at the end of E to move in again and lock the train. This cycle of events is repeated at approximately 1½ minute intervals.

In the event of the remontoire train running down the weight will continue to drive the going train and fall after the detent has unlocked the lever D. The long piece F. continues to rise until the locking piece L on its upper end contacts the lever D, thus preventing the cord being drawn off the pulley K by the outer end of G rising too high, and the clock then stops.

A similar clock, also by Breguet, may be seen in Fig. 52. Drawing by J. Martin.

Early Remontoires

Jost Burgi was possibly the first, in the second half of the 16th century, to devise a remontoire, an example being seen on the older observatory clock in Cassel. In this a weighted box is raised on a toothed rack twice a day, thus eliminating the unequal force of a driving spring.

A different approach was adopted by Nicholas Radeloff who was working in the mid-17th century. He tried to overcome the problem of providing a constant driving force with his famous rolling ball clocks in which, as the balls descend down a spiral slope they press onto bars which are linked to and drive the movement.

Harrison was one of the first in this country to make use of a remontoire to assist in precision timekeeping, employing one of 7½ seconds duration on his sea clock which performed so superbly on its' sea trials.

The period from 1750 onwards saw the emergence of several French clockmakers of superb technical skill and ingenuity such as Janvier, LeRoy, Berthoud, Motel, Lepine, Lepaute and Breguet, who devoted much of their talents to making fine watches and complex technical pieces such as table regulators. It was to these latter that they frequently applied remontoires so as to convert them from spring to weight driven, which is obviously the ideal so far as precise timekeeping is concerned, or to rewind a small spring at short intervals which then drives the clock, so as to minimize any variation in the power transmitted to the train.

It is in the field of remontoires that the French clockmakers probably displayed all their talents most effectively and it is for this reason that we are illustrating here some of the various methods they devised to overcome this problem.

Whereas the remontoire is seen most often on French clocks, Sarton in Belgium (Fig. 172.) also used this device and it was sometimes employed in England, Austria and occasionally in other countries.

Fig. 119e.

Fig. 119d.

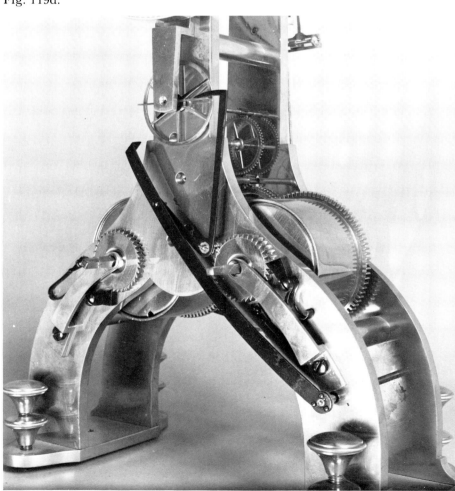

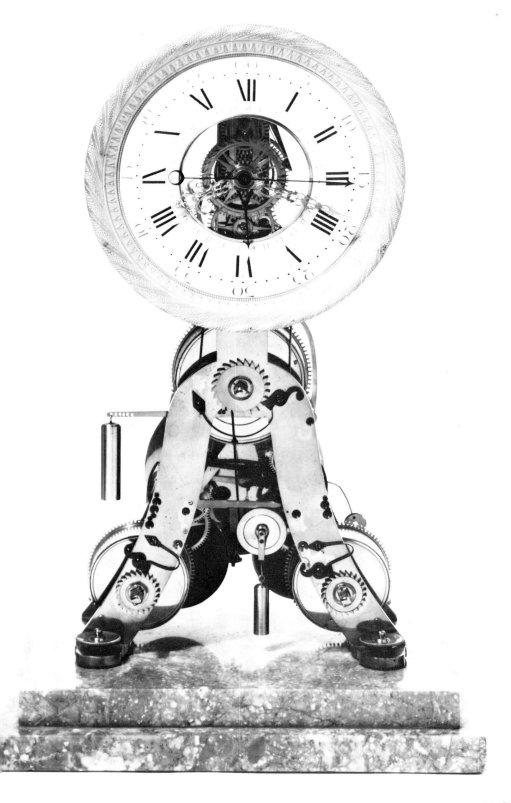

Louis XVI Table Regulator

Figs. 120a, b. An unsigned Louis XVI table regulator of two weeks duration. It chimes the quarters on two bells and strikes the hours on a third bell mounted at the bottom of the backplate. It has dead beat escapement with the anchor spanning half the great wheel, a gridiron pendulum, not shown here, centre sweep seconds and date hands and a remontoire which rewinds the driving weight every 30 seconds. Photo courtesy Keith Banham.

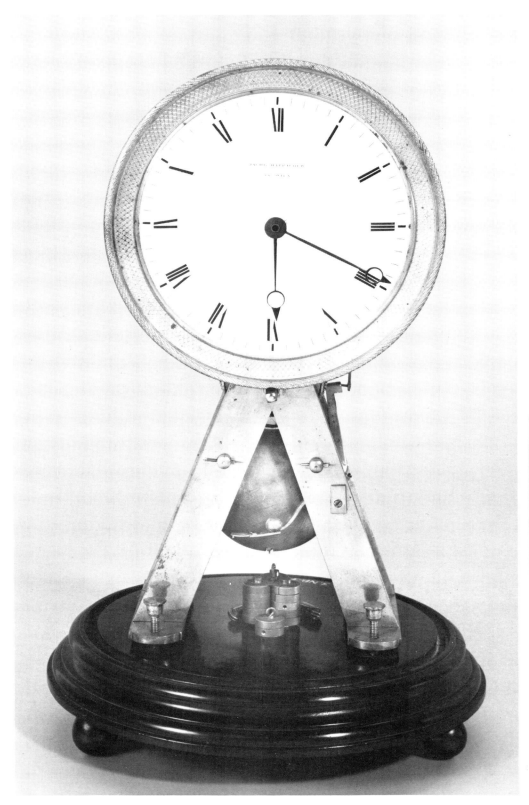

***Viennese Skeleton Clock
With Remontoire***

Figs. 121a, b. An interesting skeleton clock by that fine Viennese maker Jacob Happacher, which is in the Sobek Collection. It has a pin wheel escapement and a remontoire which rewinds the weight every few minutes, the speed with which this occurs being controlled by the skeletonised star shaped vane mounted to one side of and at right angles to the plates. When this clock was photographed the weights were not attached to the clock.

A similar but more complex clock by Bouchet of Paris may be seen in Figs. 7a and b. Photos courtesy Oster-reichisches Museum Fur Angewandte Kunst, Wein.

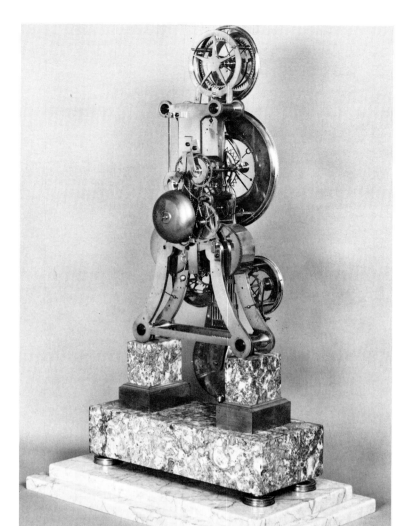

French Two Train Skeleton Clock With Remontoire

Figs. 122a, b, c. A front view of this clock and a general description are given on page 60, Fig. 46 and thus only the mechanism of the remontoire will be described here.

The remontoire train of this clock consists of the going barrel, four further wheels and the fly and is powered from a pulley wheel A mounted at the rear of the third arbor. Over this wheel runs a continuous rope of metal beads which passes under a jockey pulley B, over another pulley C of equal diameter to A and finally under another smaller pulley D which is pivotted part way along a radiused arm E. This arm is freely pivotted at its right hand end and also carries the small weight F to power the escapement.

The 'scape pinion G is driven by a wheel mounted behind the pulley C which is kept constantly rotating by the fall of the weight F. As the weight falls the arm E pivots radially about its right hand end and a small roller H, running in its left hand end, bears on the long steel hanger J, and moves it steadily to the left. This hanger is connected rigidly to a vertical piece K which has a detent at its foot and engages a locking pin on the fourth wheel of the remontoire train.

When the whole assembly has moved far enough to the left the train unlocks causing the pulley A to rotate and lift the arm E sufficiently to relock the detent on K.

The jockey pulley B is pivotted at one end of an arm which in its turn is pivotted at its other end and this assembly provides sufficient weight to keep the rope in tension throughout the cycle. Drawing by J. Martin.

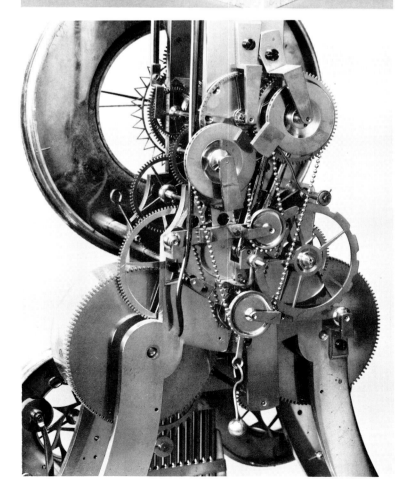

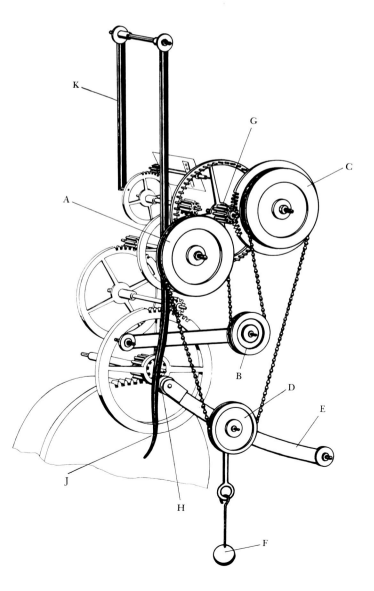

Fig. 123a.

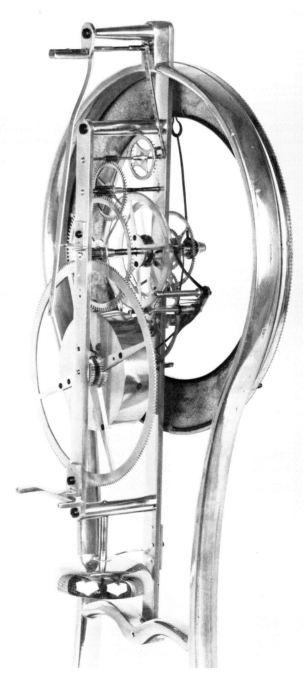

Fig. 123b.

Fig. 123c.

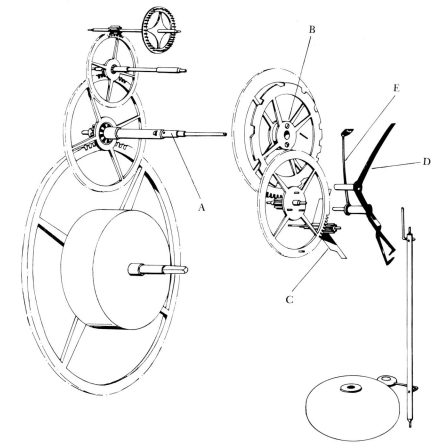

Figs. 123a, b, c, d. These keyhole shaped or extreme skeletonised clocks which were made in the first part of the 19th century frequently employ a remontoire. In most instances where this device is employed on a clock with only a single spring barrel the striking train is powered directly by the mainspring and the remontoire is used to drive the going train so as to give a "constant force" to the escapement and thus the best timekeeping. In this instance the technique is reversed, the going train gaining its' power direct from the mainspring and the strike train from the continuous rotation of the centre arbor A which winds a spring in a small barrel B on which is mounted the primary wheel of the striking train and the countwheel. The primary wheel drives a pinion carrying the second wheel and this has three lifting pins near its centre and a combined locking-off and warning pin on its rim. This wheel in turn drives a pinion carrying a three-vane fly C.

A combined warning and lifting piece D is operated in the conventional manner by a pin in the rear of the cannon wheel and by means of another pin disengages the locking piece E.

The clock employs a pin wheel escapement and strikes on a bell mounted horizontally below the dial. The long shaft for the hammer may be seen in Fig. 123c just to the right of the winding square. The pendulum is not shown in these photos. Fig. 123a courtesy R. Lister. Drawing by J. Martin.

Mid-19th Century World Time Clock

Figs. 124a, b, c. A fascinating and probably unique world time clock believed to have been made in Paris in the 1850's. The movement which consists of three separate but interlinked units, one for each train, has three massive mainspring barrels with fine steel clickwork. The outer two, for the hour and quarter strikes, are suspended on brackets below the movement plates and the going barrel runs between extensions of the mainplates. The quarter striking is unusual in that at the first quarter there is a single strike on one bell, at the second quarter one strike on each of the two bells, on the third quarter one strike on one bell and two on the other and at the hour two strikes on each bell. The hour is struck on a third bell.

This clock has a half seconds gridiron pendulum with the knife edge suspension resting on agate pads and a 60 tooth 'scape wheel with Graham jewelled dead beat pallets; power for the escapement being provided by a small spherical weight which is rewound every 5 seconds.

Fig. 128a.

Mid-19th Century World Time Clock

Figs. 124a, b, c. A fascinating and probably unique world time clock believed to have been made in Paris in the 1850's. The movement which consists of three separate but interlinked units, one for each train, has three massive mainspring barrels with fine steel clickwork. The outer two, for the hour and quarter strikes, are suspended on brackets below the movement plates and the going barrel runs between extensions of the mainplates. The quarter striking is unusual in that at the first quarter there is a single strike on one bell, at the second quarter one strike on each of the two bells, on the third quarter one strike on one bell and two on the other and at the hour two strikes on each bell. The hour is struck on a third bell.

This clock has a half seconds gridiron pendulum with the knife edge suspension resting on agate pads and a 60 tooth 'scape wheel with Graham jewelled dead beat pallets; power for the escapement being provided by a small spherical weight which is rewound every 5 seconds.

Fig. 125. The final wheel of the going train A carries a pinion B in the centre of the arbor and the speed of the train is controlled by a fly C which also carries a locking arm D. The wheel E is pivotted within a freely suspended frame F pivotting on the same centre as the 'scape pinion G and carrying an arm with an adjustable weight H.

In Fig. 125 the train is locked by the arm D. resting on an agate pallet J. The 'scape wheel is being driven by the weight H and, because the carriage F. is pivotting about the 'scape wheel centre, the wheel E. is running around the stationary pinion B. This movement finally unlocks the lever D. from the pallet and the train is released as seen in Fig. 126.

The fly is allowed to make one full turn and, in doing so, the pinion B drives the wheel E in its normal direction of motion bringing the carriage F back to the position seen in Fig. 125 with the locking piece again held by the pallet J.

The skeletonised main dial is silvered and has raised black numerals. It has a gilded bezel which is removed to provide access to the winding square for the going train. The seconds dial situated above the main dial is also skeletonised.

A single rod goes down from the movement to drive the 14 subsidiary dials which show the time in Buenos Aires, Peking, Bombay, Tahiti, Moscow, Pondicherry, Mexico, Batavia, Rio de Janeiro, Port de France, New York, Constantinople, St. Louis Senegal and San Francisco. Flanking the world time dials are two thermometers, one graduated in Reamur and Centigrade and the other in Fahrenheit and Centigrade.

The dark green flecked marble base is a replacement, the shape of every component of which was carefully copied from the original which was damaged beyond repair when the clock was dropped some eight years ago. The meticulous restoration of this clock over a period of 4-5 years took the best part of 1,000 hours of highly skilled work. Drawing by J. Martin.

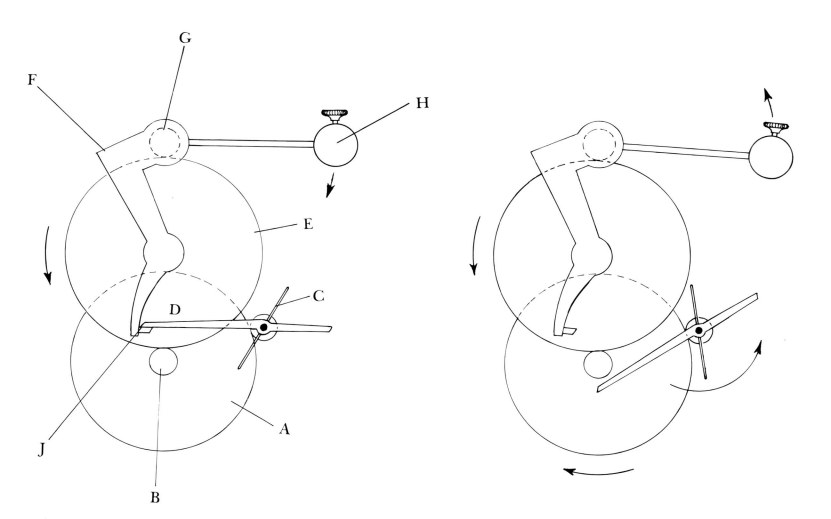

Fig. 125.

Fig. 126.

Further skeleton clocks with remontoires may be found in the French section of this book and a particularly fascinating example by Pasquale Andervelt which is powered by gas is illustrated on pages 162 and 163.

REFERENCES

Daniels, G. "The Art of Breguet", Sotheby Parke, Bernet, 1975.
Tardy, *Dictionnaire des Horlogers Francais*, 1972.
Tardy, *French Clock The World Over*, Part 2, 5th Edition, 1981, Page 171.

Table Regulator With Remontoire
By Sarton A. Liege

Figs. 127a, b, c. A general description of this clock including its' compound pendulum is contained on page 174 and therefore just the remontoire will be discussed here.

Only the striking train of this clock is directly powered by the mainspring the pulley A being squared onto an extension of the first arbor behind the backplate. Whenever the clock strikes and the train is released the driving weight B for the going train is partially wound by the pulley's rotation.

The continuous cord runs over the small pulley C which is pinned to the first arbor of the going train. This arbor carries a wheel with 72 teeth engaging a pinion of 6 followed by a wheel of 60 which engages the 'scape pinion of 6. This escapement employs a pin wheel D which has 30 teeth on each side. The driving cord is kept tensioned by the counter weight E. Drawing by J. Martin.

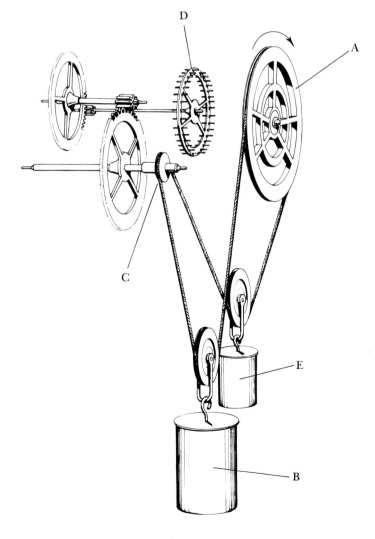

Fig. 128a.

3

Austria

The classification "South German Clocks", which is generally used to refer to those made in this area in the 15th-17th centuries, has in the past usually included Austrian clocks. However, even at this early period clocks made in the capitals of the old monarchies of Vienna, Prague, Innsbruck, Graz and other smaller towns, had many similar features which enables them to be grouped under a separate heading, and it is in this group that many of the finest and most inventive clockmakers such as Emmoser, clockmaker to Maximilian II and Burgi of The Prague School who invented the cross beat, are contained.

It must always be remembered that Austria was part of the Austro-Hungarian Empire, which lasted from 1520-1918 and that it was only after this that Austria became an entirely separate country. It is largely for this reason that clocks made in Austria, Hungary and Czechoslovakia, for instance, often bear such a close similarity.

The invention of the pendulum in 1657, together with the internal changes taking place in Europe at that time, led to the demise of the South German clockmaking industry and marked the beginning of the dominance of English clockmaking, the influence of which was to spread throughout Europe. Many English clocks were imported into Austria and the English style of bracket clock was rapidly copied, although in a slightly more flamboyant style. There was also a considerable intermix between the two countries. Thus not only were English clocks exported to Austria, but clocks are seen which appear to be Austrian yet have typically English movements. It is likely that in these cases sometimes the movements were exported to Austria, either complete or in an unfinished state, or that on some occasions they were made by English clockmakers living in Vienna.

It is interesting that Dowe Williamson emigrated from England to Austria in 1721, and was given a franchise to produce complex clocks, and in the 1740's the University of Vienna held seminars dealing with complex mechanisms. Amongst those who attended was David Cajetano who later produced his celebrated astronomical clock.

When considering Austrian clockmaking it should be kept in mind that Austria has always been a buffer state between East and West, and has been influenced by many different countries, whose crafts and sciences it has rapidly assimilated and expressed in its own way. Austria itself, on occasion, has also been a great center of cultural excellence which was undoubtedly stimulated by the fact that it was the most influential of all the Imperial Courts in the Austro-Hungarian Empire.

In the renaissance period clockmakers in Austria belonged to the locksmiths guild and were basically divided into two groups, Grossuhrmachers (big clockmakers) who made turret, weight driven and wall clocks, and Kleinuhrmachers (small clockmakers). This latter group originated from clockmakers who specialized in the manufacture of watches following their invention in the 16th century, but who also made small clocks which were usually spring driven.

Figs. 128a, b. This fine quality clock which is signed on both the front and back of the inverted Y frames by Franz Denk in Wien, is most unusual for an Austrian clock of this period, in that it has a chain fuzee, although this was relatively common for the going train of Austrian table and mantel clocks made at an earlier period.

It has an enamelled chapter ring and a delicate pin wheel escapement. The upper end of the pendulum, which has a large brass bob, is fretted out in the form of a shield and rests on a knife edge. The wheelwork is quite exceptional with every wheel being crossed out in a different and highly individual style. Height: 11¾" (26 cms.). Collection Dr. Sobek. Photo courtesy Osterreichisches Museum Fur Angewandte Kunst, Wein. Photography by Norbutt Lieven.

Fig. 129. An eight day Austrian skeleton timepiece with inverted Y frame, anchor escapement and a decorative steel rod pendulum. Height:10⅝"(27 cms.). Photo courtesy Museem der Stadt Wien.

In the late 17th and the first half of the 18th century, the influence of English clockmaking was very strong, the bracket clocks for instance looking very similar to those produced in England, however, by the 1780's this influence had started to wain.

To improve the standard and efficiency of their watch and clockmaking, Swiss craftsmen were invited in 1779-80 to work in Vienna where workshops and accommodations were provided for them.

Initially there were some fifty in this group, but their number eventually rose to one-hundred and fifty. Amongst them were many specialist craftsmen such as handmakers, engravers, gilders, file makers, etc. It seems extremely likely that these formed the basis of the skills which enabled the production of many fine clocks over the next one-hundred years.

The Vienna Empire and Biedermeier Period

By the end of the 18th century, the influence of the French was beginning to be felt quite strongly. Napoleon arrived in Vienna in 1804 and married Marie Louise, the daughter of the House of Hapsburg, Austria's ruling family, and the Empire period was now born, inspired by the conquest of Italy. This gave rise to a return to classical proportions and subjects, and resulted in the birth of the "Vienna" regulator, one of the most attractive clocks ever conceived, with excellent lines and, particularly in the early stages, great delicacy of construction. It also gave rise to clocks based on classical themes such as Roman temples and statues. However, whereas in France these were nearly always made in fire gilt bronze in Austria the majority were, probably largely for economic reasons, made of gilt wood.

Under the influence of the Swiss and French, Austrian clockmaking now changed quite dramatically and gave birth to a new generation of clocks never really seen before in Europe. Above all else, it was their delicacy that gave them much of their appeal, contrasting so strongly with the fine quality, but usually massive movements being made in England, and to a slightly lesser extent in France.

The years 1790-1860 could well be called the golden era of Austrian clockmaking. The ingenuity of the clockmakers at that period was very prodigious as can be seen from some of the clocks illustrated. They were produced with complex pendulums, escapements, and calendar work and were frequently of long duration, running for a year and sometimes longer.

The Empire Period extended from 1800-1820, and gradually gave way to the Biedermeier, which really marked the birth of the industrial revolution and from 1820-1870 the number of clockmakers employed in Vienna steadily increased. However, as the end of the century approached Viennese clockmaking gradually declined because it was unable to meet the competition of the mass production of similar clocks in Southern Germany and the large quantities of cheap clocks being exported from America and France.

REFERENCES

Claterbos, F.H. von Weijdom, *Viennese Clockmakers and What They Left Us,* Interbok International B.V., Schiedam, Holland.

Hellick, E., "Alt-Wiener Uhren", *Die Sammlung Sobek im Geymuller-Schlossl, 1750-1900,* Verlag Georg D.W. Callwey, München, 1978.

von Bertele, Prof. Dr. Hans, *Osterreichische Uhren Nech einem Vortrag bei der Wiener Gesellschaft der Museums freunde im Fruhjahr,* 1957.

Virtually all of the skeleton clocks illustrated in this section come from the Empire and Biedermeier period, and whereas some of these bear a superficial resemblance to French clocks produced at the same time, they are almost invariably more "airy" and delicately constructed and in the majority of cases are not made with such serious horological intent as the products of Janvier, Lepaute, Lepine, Breguet and other fine French makers. Indeed, in many instances the clocks appear to have been created almost as an effervescent display of the clockmakers art to give joy to both the creator and the beholder. The fascinating technical details incorporated in them may sometimes have little scientific basis, but do indeed make them most intriguing to look at. However, it must not be conceived from this that no fine technical clocks were made in Austria; some of the Vienna Regulators, for instance, were built to a very high standard and performed with a great degree of accuracy, and the table regulator by Pechan is as fine as any produced in France. In Austria, skeleton clock manufature probably commenced in earnest a little earlier than in England, but as with England the first clocks were quite strongly influenced by those being made in France. Thus that seen in Fig. 128 could at first glance easily be mistaken for one made in that country with its typical inverted Y frame and pin wheel escapement. Similar remarks could probably apply equally well to the one shown in Fig. 129. However, the Austrian skeleton clock rapidly assumed a character all its own, and at that stage could certainly not be mistaken for the product of any other country, except possibly its neighbors within the Austro-Hungarian Empire. For convenience we have divided Austrian skeleton clocks into four groups:

a) Spring driven clocks
b) Weight driven clocks
c) Complex clocks
d) Skeletonized clocks

SPRING DRIVEN CLOCKS

Fig. 130. Fire gilt skeleton clock by Wibral, one of the best of Viennese clockmakers, a fascinating piece by him being illustrated in Fig. 155. It has three going barrels with 4/4 strike on two gongs. The brass rod pendulum with brass bob has knife edge suspension and pin-wheel escapement. Height: 17" (44.2 cms.) Collection Prof. Hans von Bertele

Fig. 131. A most decorative
Viennese clock with columns
covered in beadwork in the
form of Petit-point embroid-
ery, a feature which almost
epitomizes the lighthearted
and gay attitude which the
Austrians seemed to have
taken to much of their clockmaking at that time and
conjures up pictures of the
lively, vivacious city Vienna
must have been. The clock is
of one day duration with 4/4
strike on two gongs and has
anchor escapement with silk
suspension. The auxiliary
dial in the center indicates
the days of the month. It is
signed on the dial Vonder-
heid in Wien and would date
circa 1830-40. Height: 17¼"
(44.8 cms.) Collection Dr.
Sobek. Photo courtesy
Osterreichisches Museum
Angewandte Kunst Wien.
Photograph by Norbutt-
Lieven.

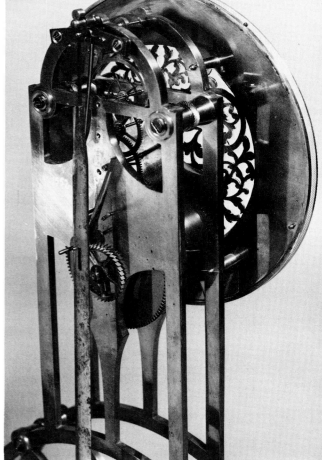

Figs. 132a, b. A *month* duration spring driven skeleton clock signed on the silvered chapter ring which has a gilded filigree center Dobner in Wien. Circa 1830-40. The frames are far more substantial than on most Austrian skeleton clocks. It has good five spoke wheelwork and a dead beat escapement with a delicate escape wheel and pallets spanning twelve teeth. Height: 21½'' (55 cms.). Collection Dr. Sobek. Photo courtesy Osterreichisches Museum Fur Angewandte Kunst Wien. Photograph by Norbutt-Lieven.

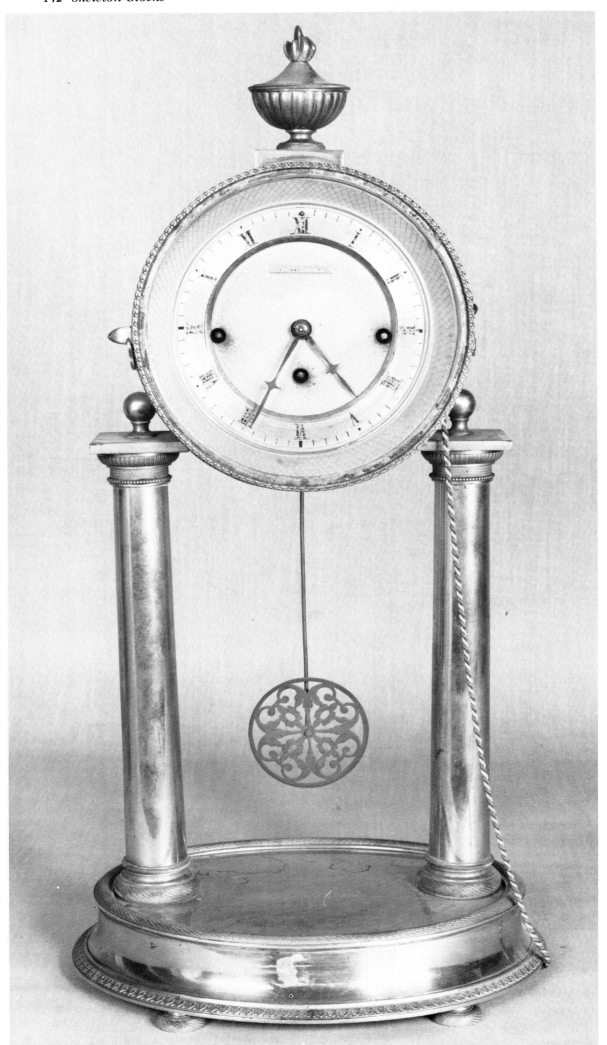

Fig. 133. A one day pillar clock with skeletonized movement; 4/4 strike and repeat on two gongs and strike/silent regulation. It is signed by Anton Liszt, the brother of the famous composer. Circa 1840. Note the delicately fretted out pendulum bob which was popular at that time. Collection Prof. Hans von Bertele.

Fig. 134. A forty hour skeleton clock with 4/4 strike and repeat, a silvered case and an elaborately pierced out pendulum bob. Photo courtesy Museen der Stadt Wien.

Fig. 135. A Viennese quarter striking skeleton clock of forty hours duration with fully skeletonized movement, dial and pendulum bob and silver plated columns. Height: 16'' (41 cms.)

Fig. 136. A decorative skeleton clock circa 1860, with silvered and gilded case decorated with garlands of flowers and classical scenes. The forty hour movement has hour and quarter hour strike and repeat on two gongs and music in the base. Photos Courtesy Museen der Stadt Wien.

Fig. 137. An Austrian mid-19th century skeleton clock which epitomizes the gaiety and fun of Vienna at that time with its ornate wood frame decorated with gold leaf and smothered with silver repousse work. That on the base incorporates figures, a cornucopia of flowers and floral swags. There is silver floral strap work curling up the pillars and behind the pendulum, the bob of which is fretted out and decorated with colored stones, behind it is a harp.

The silvered brass dial has Roman numerals and a fretted out brass center. The ornate silvered bezel incorporates a cherub. The thirty hour movement strikes the halves and hours on a gong and there is a music box in the base which plays one of two tunes sequentially either on the hour or when the repeat cord is pulled. Height: 22" (55.3 cm).

Fig. 138. A forty hour skeleton clock quarter chiming on two gongs, which bears the signature of H. Kerohnawe. Although the dial has been fully skeletonized the movement plates have been left solid. The bezel is engine turned and gilt, a feature popular on many Viennese clocks at this time. The pendulum bob is attractively fretted out and the decorative effect is increased by the gilt lyre placed behind it. Height: 20'' (50.8 cms.). Collection Prof. Hans von Bertele

Fig. 140. We are grateful to the Time Museum for a photo and description of this unusual, unsigned Austrian clock of month duration with constant force escapement, the impulse being derived from a small cylindrical weight having a one minute remontoire. It has Viennese strike, i.e. the same as French Grande Sonnerie but with the hours being struck after the quarters instead of before. The compound pendulum has double knife edge suspension and beats half seconds. Height including base: 15¼" (38.7 cms.). For a fuller description of this clock see W. Pinder's article in *Antique Horology* Dec. 1967. "A constant force clock with Grande-Sonnerie striking".

Fig. 139. A good quality Viennese skeleton clock of forty hours duration which, most unusually, is supported by three instead of the usual two pillars. The movement is partly and the dial fully skeletonized and it has quarter and hour strike and repeat on two gongs. Photo courtesy Museen der Stadt, Wien.

Figs. 141a, b. An
interesting little
Austrian skeleton
timepiece with the
lever escapement
employing a large
compensated balance
mounted vertically
above the movement.
Photos Courtesy
Museen der Stadt,
Wien.

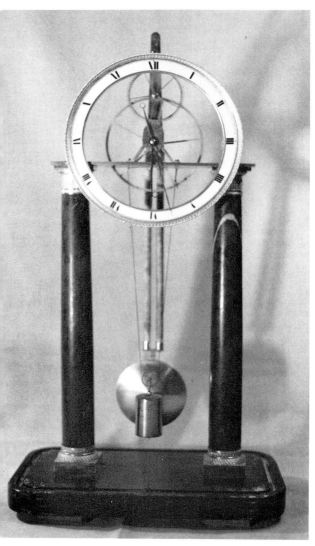

Fig. 142. A weight driven timepiece of one day duration with twin glass rod and mercury pendulum. It has only two wheels, a great and an escape wheel. Circa 1820. Collection Norman Langmaid.

Figs. 143a, b. An anonymous early 19th century weight driven skeleton timepiece of one day duration. The pendulum rod incorporates two glass tubes containing mercury and the dead beat escapement has an anchor spanning twenty-five teeth. Height: 22" (56 cms.). Collection Dr. Sobek. Photo courtesy Osterreichisches Museum Fur Angewandte Kunst Wien. Photograph by Norbutt-Lieven.

WEIGHT DRIVEN CLOCKS

English weight driven table clocks are very rare and even in France they were only made in limited numbers and generally in the form of fine quality weight driven table regulators, some of which were skeletonized. By contrast in Austria they were produced quite frequently. The reasons for this are not really clear. Although theoretically a weight will provide a more constant driving force than a spring and thus produce better timekeeping, it is doubtful if this consideration would apply to most of the clocks in question. A more probable explanation is that it made the clocks more attractive and interesting to look at which is also the reason why most people prefer a weight to a spring driven Vienna regulator.

The simplest of the weight driven clocks such as those seen in Figs. 142 and 143 and which are of one day duration, have only two wheels, a great and a 'scape wheel, and even these are crossed out as delicately as possible so that it appears as if the clock has practically no movement. At the same time the chapter ring is reduced to the minimum width, particularly on the earlier clocks, adding still further to the appearance of delicacy. Twin glass tubes containing mercury were quite frequently used for the pendulum rod. However, it is unlikely that they provided any accurate compensation. With the earlier clocks the pendulum bob tended to be solid but on the later examples (Fig. 138), it was often attractively fretted out.

To try and get the maximum fall for the weight, particularly on the eight and thirty day clocks, there was often a circular cut out in the base into which the weight could drop. In the majority of cases the weights were circular and left plain, but on occasions they were rectangular, particularly on the three train clocks and sometimes they were engraved.

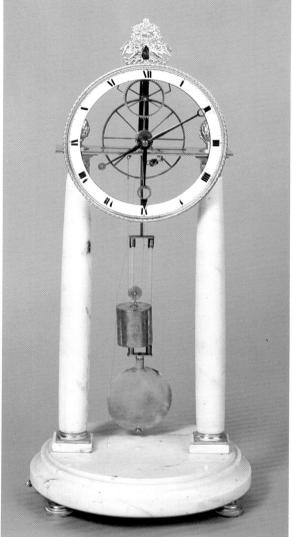

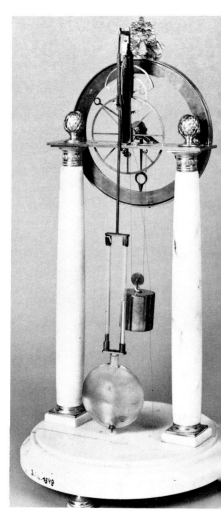

Fig. 144. A fine quality eight day timepiece with very delicate hands and a beautifully crossed out great wheel with inner and outer rims connected by twelve spokes. It is signed on the backplate by the eminent Viennese maker Wibral and numbered ten which would give a date of manufacture of around 1805 as Wibral was working from 1804-23. Collection Prof. Hans von Bertele.

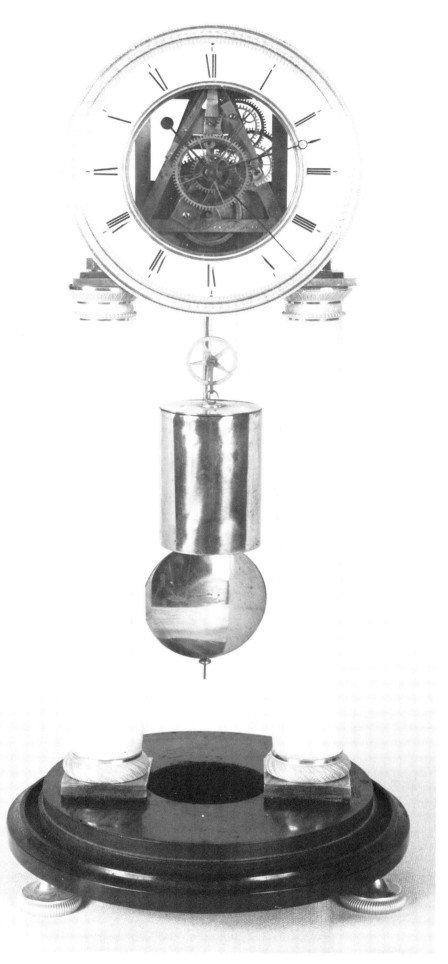

Figs. 145a, b. An unusual weight driven eight day time-piece signed along the base of the front frame Joseph Kostler, Wien who was working around 1814. It has fine wheelwork and a center sweep seconds hand which, because a coup perdu escapement is used in conjunction with a half seconds pendulum, shows true seconds. The columns are alabaster, the capitals fire gilt and there is a hole in the ebonized base into which the weight drops. Collection Norman Langamaid.

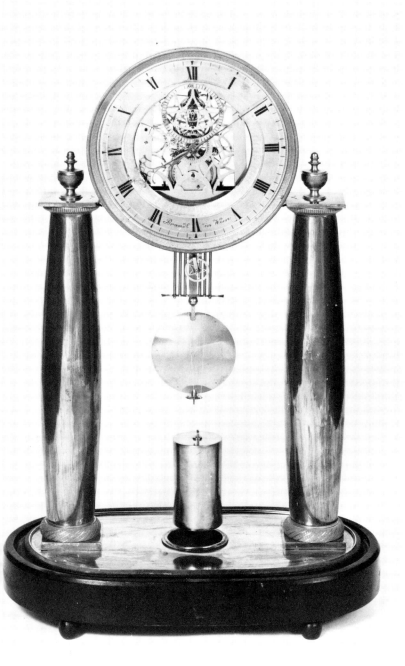

Figs. 146a, b. A fine quality 19th century month-duration weight driven timepiece. A nice refinement is the adjustment for pallet depth. Height: 19¼'' (49 cms.). Collection Dr. Sobek. Photo courtesy Osterreichisches Museum Fur Angewandt Kunst Wien. Photography by Norbutt-Lieven.

Fig. 148. A very decorative *month duration* weight driven skeleton clock signed on the front frame Jos. Binder in Wien, 1823, probably the date of manufacture. It has a pin wheel escapement, knife edge suspension, maintaining power and a seconds dial below twelve o'clock on the silvered chapter ring. A nice refinement is the adjustment for pallet depth. Height: 19¼'' (49 cms.). Collection Dr. Sobek. Photo courtesy Osterreichisches Museum Fur Angewandte kunst Wien. Photography by Norbutt-Lieven

Fig. 147. A most attractive Viennese skeleton clock circa 1820-30 of eight day duration with a fine driving weight decorated with engine turning.

The clock has an engraved and silvered brass chapter ring with inner and outer engine turned bezels. Through the center may be seen the delicately skeletonized movement with beautifully shaped knife edge suspension to the pendulum which has a glazed bob in the center of which are displayed the pivoted levers for Ellicott's form of compensation.

The movement is supported by two Bohemian cut glass columns with red overlay and ormolu mounts which rest on a substantial clear glass base. Height: 22½'' (57.2 cms.). Photo courtesy Fanelli Antique Timepieces Ltd., New York.

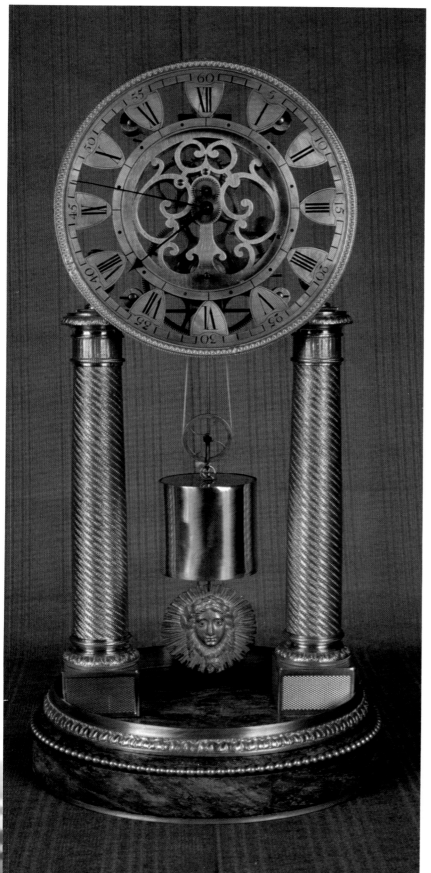

Figs. 149a, b, c. Another month duration timepiece by Binder which is attractively and very fully skeletonized. It has dead beat escapement; maintaining power and the columns, base and pendulum bob are all finely engine turned and fire gilt. The hands are very delicate and there is a hole in the green marble base for the weight. Height: 15'' (38 cms.). Collection Norman Langmaid.

Figs. 150a, b. An Austrian thirty day skeleton timepiece in which virtually the entire clock except the actual movement and pendulum bob is silvered. It has dead beat escapement with beat regulation; maintaining power and a 5 lb. engine turned driving weight which passes through the base of the clock. The front and back plates are unusual in that they are pierced out in the form of the initials, TH or HT, presumably those of the clockmaker.

The dial has center sweep minute and hour hands, seconds at twelve o'clock, the hand making one revolution every thirty-two seconds, days at four o'clock and months at eight o'clock. Height: 21'' (53.4 cms.). Formerly in the collection of Norman Langmaid.

Figs. 151a, b. An early 19th century weight driven eight day skeleton clock very similar to that seen in Fig. 153, with quarter strike on two bells and a third bell for the hour. It has an enamelled chapter ring bearing the name of Berlinger. The dark green stone frame and base supporting the movement are replacements as also are the weights and the glass tube mercury compensated pendulum, the bob of which bears the insignia of Prof. Hans von Bertele in whose possession this clock resided until his death in 1984.

Figs. 152a, b, c. A highly individual Austrian eight day skeleton clock signed on the dial and backplate Peter Goetz in Wien. The 2½ plate movement has a skeletonized backplate with a weight driven going train which is provided with stop-work and deadbeat escapement and spring driven Viennese 4/4 strike and repeat on two gongs with provision for strike/silent regulation.

The dial layout is most unusual and at first sight appears similar to that of an English regulator. However, whereas seconds are still shown below twelve o'clock, the hours and minutes are transposed. The counterpoised center sweep hand, which has a spring loaded clutch so that it can be moved out of the way during winding, indicates hours with each quarter hour being numbered, i.e. 1, 2, 3, and the lower dial gives minutes. This is marked I, II, III, and IIII, again to signify the quarter hours, but is also marked out in minutes although these are not numbered. The dial is surrounded by delicate silver fretwork and the bob is also attractively pierced out. Height: 17½'' (44.5 cms.)

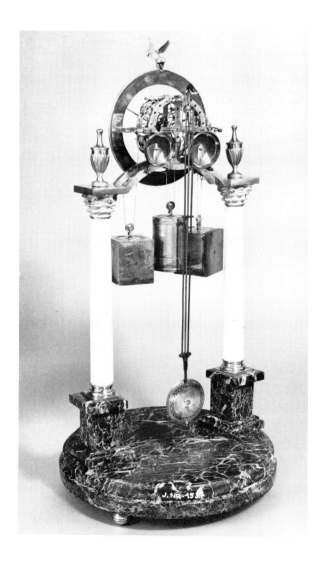

Figs. 153a, b, c. An anonymous eight day skeleton time-piece with silvered brass chapter ring, striking the quarters on two bells with a third bell for the hour. It has a Brocot escapement with adjustable steel pallets; a mock compensated pendulum with a face depicted on the bob and engraved weights.

The frame is in the form of a ring with a cross, as is that of the clock seen in Fig. 151 by Berlinger. The quarter and hour striking trains, both of which have very delicate wheelwork, are arranged around the periphery and the going train runs up vertically in the center. Even the star wheel and flies on this clock are skeletonized. Collection Dr. Sobek. Photo courtesy Osterreichisches Museum Fur Angewandte Kunst Wien. Photography by Norbutt-Lieven.

COMPLEX CLOCKS

The clocks contained in this section have been chosen to illustrate the extreme ingenuity, individuality and quality of workmanship of the best Austrian clockmakers. Whilst some are obviously inspired by clockmaking in France at that time they are in no sense copies, the beautiful table regulator by Pechan, for instance, although incorporating many technical details originating from France, is entirely different in its overall concept, and many would consider it more attractive, containing great delicacy of design.

The novel concept of the clock by Brauner seen in Fig. 154, and the artistry of its design and execution, are typical of the finest Viennese work, which by this time had assumed a character all its own. Nor were the Austrians averse to producing technically complex clocks such as those seen in Fig. 157 and 161. However, here they differed from the French and English, in that the artistry and overall appeal of the clock was of paramount importance, frequently, for instance, an attractively designed and apparently compensated pendulum was in fact no such thing; whereas in England where quality and integrity and the best possible performance, occasionally achieved almost regardless of appearance, was the dominant influence, no such similar light hearted approach would have been tolerated.

Fig. 154. A beautifully designed and executed early 19th century fire gilt timepiece by the eminent maker Josef Brauner in Wien who is recorded as working from 1804-1828. At first sight it appears to have neither a weight, nor a spring to drive it but on closer examination a fine line can be seen running down into the base where a spring barrel is concealed.

The hands are very fine and the movement is well executed with dead beat escapement with adjustable pallets; maintaining power and end screws to the upper part of the train. The pendulum rests on a knife edge. Overall height: 15'' (38 cms.). Collection Prof. Hans von Bertele.

In France the finest technical pieces produced by some of their best makers such as Janvier, Lepaute, Lepine and Breguet to name but a few obviously exceeded in quality almost everything made in Austria. Nevertheless, they must have been many times more expensive and did not necessarily look more attractive as they lacked the delicacy of the Austrian clocks. However, to behold the best French gridiron pendulums, for instance, is almost awe inspiring in that it was something conceived and executed almost regardless of expense and produced virtually nowhere else in the world. It is even more impressive when one remembers that the same or better results could have been achieved by the much simpler pendulums being used in England at that time.

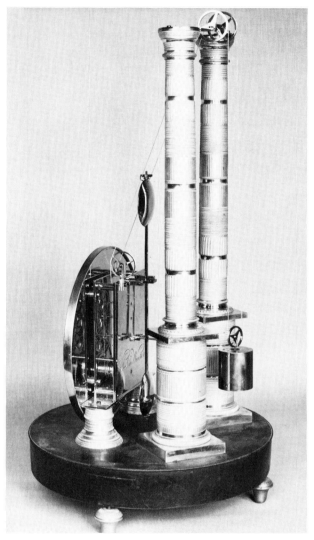

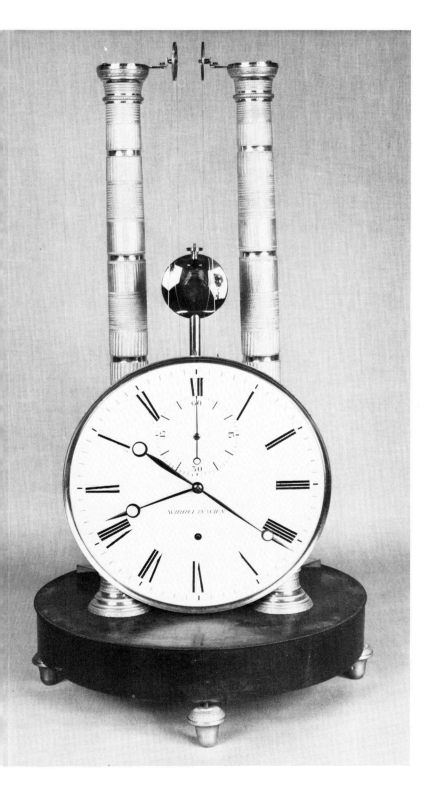

Figs. 155a, b. This fascinating weight driven timepiece which was in the collection of Prof. Hans von Bertele, is discussed in Bassermann-Jordan/Bertele pp. 462, 465, 466. When purchased it was without any traces of the pendulum suspension and over a period of thirteen years failed to run happily no matter what was tried until a similar clock by Janvier with point suspension was seen and copied. Thereafter the clock went regularly with the original light weight.

It is signed on both the enamelled dial and the backplate Wibral in Wien and would date circa 1810. It has a seconds beating compound pendulum with beat regulation and a pin wheel escapement. The line for the driving weight runs up from the movement to the top of one of the beautifully finished gilded pillars, thence down to the weight pulley, back to the top of the other pillar and down to the movement where it is tied off. Mathias Wibral, a fine maker, was working from 1804-1823.

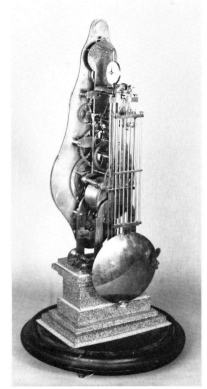

Figs. 156a, b, c. This very fine skeletonized table regulator was made at the end of the 18th century by Pechan of Vienna for Emperor Franz I of Austria, and is believed to be shown with him in a painting by Peter Fendi. It was subsequently given by the Emperor to one of his ministers.

The clock is of month duration, has a coup perdu pin-wheel escapement and a complex compensated gridiron pendulum. A seven rod gridiron bar rises up from the base and supports a bracket which extends backwards to carry the relatively conventional seven rod gridiron pendulum which is suspended by a steel strip from a pivoted bar. Fast/slow regulation is achieved by moving the steel hand on the enamelled dial which acts on the other end of the pivoted lever and raises or lowers the pendulum. Because the suspension strip passes through a slit in a brass block, this in effect varies the pivot point and thus the effective length of the pendulum. Once correct regulation has been achieved the suspension can, if desired, be clamped. Two pivoted clamps are provided for the pendulum bob and beat regulation is also fitted.

The gilded brass dial plate has the following dials on it, going from top to bottom.

1. An enamelled dial for the years marked 1-4 thus giving leap year.
2. A two handed enamelled dial showing the days of the month and the months of the year.
3. An engine turned gilt dial with an outer narrow raised and silvered chapter ring with a beautifully balanced minute hand; there is another chapter for hours below twelve o'clock and a third at six o'clock for seconds.
4. A world time dial, the outer enamelled chapter with Vienna at the top bearing the names of forty-eight cities throughout the world and indicating the time in these places by means of an inner rotating enamelled disc numbered 1-24 in Arabic. Within this is a rotating lunar disc and in the very center a four sided silver star, around which is written Siden, Osten, Norden, Westn (N.S.E.W.), thus indicating the relative position of the moon in the heavens.
5. At the bottom is the winding square with a decorative silver arrow, indicating the direction of wind and at the periphery a hand shows the state of wind against a scale graduated 1-31. Collection Prof. Hans von Bertele.

Figs. 157a, b, c. One of the most ingenious clocks it has been our pleasure to examine, which would have been made around 1812. The fourteen day movement has Viennese quarter strike on two gongs with three going barrels, pin wheel escapement, steel strip suspension and an enamelled dial with a steel hand built into the bob for fast/slow regulation. The enamelled chapter ring bears the signature Brandl in Wien, an emminent maker.

There are three hands and five miniature dials which rotate on them and are operated by trains behind the dials connected to revolving weights.

The dial in the tip of the minute hand shows hours, whereas the one in its tail shows minutes. The hour hand carries a spherical moon (which does not rotate) in its pointer and indicates days of the week on the dial in its tail and the calendar hand, which shows the day of the month against the inside of the chapter ring, has a subsidiary dial in its tail for months of the year. Collection Dr. Sobek. Photos courtesy Osterreichisches Museum Fur Angewandte Kunst Wien.

Pasquale Andevalt's Hydrogen Clock

Pasquale Andervalt, who lived in Trieste from 1806-1881, which was at that time part of the Austro-Hungarian Empire, devised a most ingenious remontoire powered by hydrogen which was released when pellets of zinc came in contact with sulfuric acid.

The gas gradually forces up a bell shaped container which takes up with it the carriage, and thus raises the driving weight.

Several of these clocks exist. One is in the collection of the Worshipful Company of Clockmakers (Fig. 158) catalogue No. 518 (Catalogue printed 1949), another is in the Vienna Clock Museum, and a third came up for sale in Sothebys in February 1986.

REFERENCES

Bassermann-Jordan/Bertele, *The Book of Old Clocks and Watches, 4th Edition*, Translated from the German by H. Alan Lloyd, 1964, pp. 472.
Bruza, Giuseppe, *L'Arte dell Orologerie in Europe*, p. 159, Note 301.
Morpurgo, Enrico, *Orologiai Triestini*, Le Classidra Anno IX, Nov. 1953, N. 11, pp. 5-7.

Figs. 158a, b. This example, which is in the collection of The Worshipful Company of Clockmakers, is similar in most respects to that seen in Figs. 162a, b. but has a spiral tube mounted above it containing the pellets of zinc which are released when the jar containing the gas reaches the top of the container. Photo Courtesy The Worshipful Company of Clockmakers.

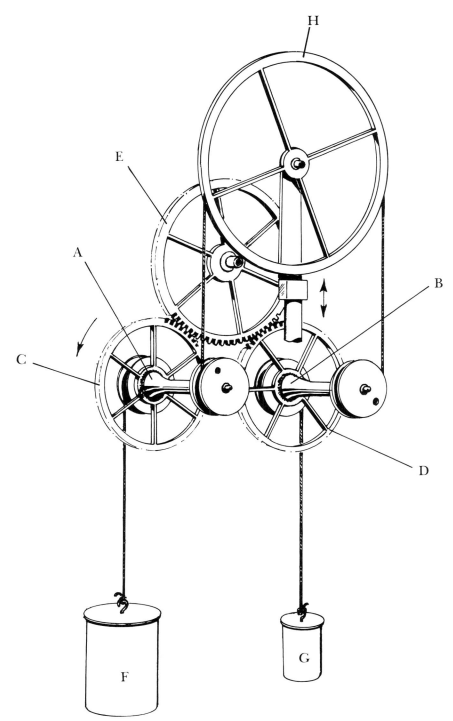

Fig. 159. *The Gas Powered Remontoire devised by Pasquale Andervalt, Trieste.*

Two spools, a and b, are pivoted between the plates. The spools both carry a barrel at each end and wheels c and d are mounted freely behind the forward barrels and are each held by conventional click-work. Both wheels engage a larger wheel e, from which the drive to the train is taken.

The driving weight f hangs from the forward barrel of spool a. A counterweight g hangs from the forward barrel of spool b. A continuous line from each of the rear barrels runs over a large pulley wheel h. This latter is freely pivotted on an extension to the piston which operates vertically in the glass gas container.

Operation

As the driving weight f falls, the wheel c maintains a positive drive on the wheel e. The rotation of the spool a wraps the cord running over the pulley h around the rear barrel, thus maintaining a positive drive to the wheel e through the other spool and also winds the counterweight. While this is happening the pulley h is slowly descending as the gas pressure diminishes. On reaching the lowest point of its travel a further zinc pellet is dropped into the acid in the gas container causing the pressure to increase, and the pulley h and the piston assembly to rise slowly upwards.

This upward movement rotates the driving spool a in the reverse direction, thus rewinding the drive weight. It does, however, prevent the wheel c from driving, but this function is taken over by the wheel d as, while the large pulley rises it applies power to the spool b, thus maintaining the positive drive to the wheel e.

This arrangement of maintaining power is essential, as the action of a remontoire, dependant on a chemical reaction, is necessarily slow and it must take some time for the pulley h to reach its upper position when the driving weight will be fully rewound and can recommence to apply a positive drive. Drawing by J. Martin.

Fig. 160. A fully skeletonized temple clock with a rotating band, indicating the time against the head of a stationary snake. The one day movement has rack striking for the hours and quarters with three going barrels. The 'scape and contrate wheels may be seen just above the inverted bell which rests on an attractively fretted out platform. The seven pillars represent the days of the week and the four caryatids, the four seasons. Collection Dr. Sobek. Photos courtesy Osterreichisches Museum Fur Angewandte Kunst, Wien.

Fig. 161. A skeleton clock of highly individual design made by Alois Schenk of Vienna around 1840, and signed by him on the front plate. The eight day movement has a pin wheel escapement, knife edge suspension and count wheel strike on a gong mounted at the base of the clock. It has a single large going barrel which provides power directly for the striking train and indirectly for the going train through a weight driven remontoire, much of the mechanism of which can be seen on the front frame. The dial layout is possibly unique. There are "fly back" hands registering against tapering vertical enamelled scales; that on the left giving minutes and the one on the right hours. Every time the minute hand falls down to the bottom of the scale it releases the strike. The very narrow chapter ring in the center is for seconds and the small dial above it is for hand-set (Richte die Zieger). Much of the wheelwork is crossed out in the form of the "Star of David". Photo courtesy Museen der Stadt, Wien.

Figs. 162a, b. Hydrogen clock, the mock compensated pendulum signed Pasquale Andervalt I.R. Priv. E. Prem Fabbricatore in Trieste, bears the enamelled Imperial Arms of Austria and is overlaid with gilt leaf scrolls. The movement, which has lyre shaped plates and a pin-wheel escapement, is numbered 101. In this example it is believed that a hollow zinc rod was probably fitted below the central brass rod whereas on the clocks in the Museum in Vienna and the collection of The Worshipful Company of Clockmakers pellets of zinc were used.

Ref: Sotheby's Auction Catalogue 20.2.86. Photo Courtesy Sotheby's.

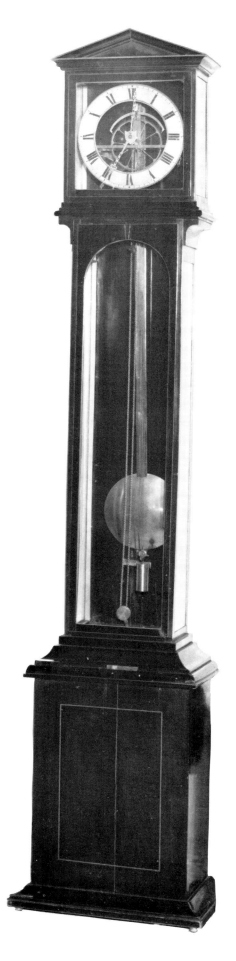

SKELETONISED CLOCKS

Some of the finest "Vienna Regulators", particularly of the floor standing variety and those containing rare or unusual features, were skeletonized so as to display the clockmakers skill and ingenuity as completely as possible. In this section we have chosen four longcase regulators and one miniature wall clock to give some idea of the almost infinite variety of the movements being produced in Vienna, mostly in the late 18th and the first part of the 19th century.

Figs. 163a, b. An unsigned pearwood longcase regulator of one day duration, with Huygens endless cord winding and a steel rod pendulum with large brass bob. As can be seen it has no train, just one very large 150 tooth scape wheel which revolves once every five minutes, and is divided by the spokes into ten sections. It is the only escape wheel which the author can remember on which seconds are recorded, the sectors at the base of each spoke being marked out alternately thirty and sixty with six divisions, each representing five seconds between. There is a pointer below twelve o'clock to indicate the time in seconds and a massive anchor spanning fifty-six teeth which has very delicate dead beat pallets. Height: 6' 1¾" (2.13m). Collection Dr. Sobek. Photo courtesy Osterreichisches Museum Fur Angewandte Kunst, Wien.

Figs. 164a, b. A rosewood longcase regulator with fully skeletonized movement made by Marenzeller of Vienna in the second half of the 19th century. The month duration movement, which has wheels of very high count, has hour and half hour strike on coiled gongs, dead beat escapement with adjustable pallet stones and a pendulum consisting of three glass tubes (probably not original) with a brass cased bob.

An unusual feature is the central calender dial on which the hand driven by two beautifully skeletonized star wheels turns once every thirteen weeks i.e. four times a year. This is achieved by a pin on the hour wheel engaging the smaller and lower fourteen tooth star wheel twice a day and thus making it complete one revolution a week. This star wheel in turn has a pin on it which meshes with the larger star wheel once a week and thus advances the calender hand by the appropriate amount. Collection Dr. Sobek. Photos courtesy Osterreichisches Museum Fur Angewandte Kunst, Wien.

Figs. 165a, b. A very complex early 19th century eight day skeletonized longcase regulator with seconds beating nine rod gridiron pendulum with knife edge suspension and pin wheel escapement, signed on both the dial and frontplate Josef Ettl in Wien.

This clock has two striking trains. One, which is weight driven using Huygens endless cord, strikes the solar hours and quarters using one hammer and two bells and the other is powered by a spring, and has mean time Viennese 4 4 strike on two gongs. (Which may be seen in the photo above the year calender.)

The rope wound going train is weight driven and shows both mean and solar time. There are five hands; two of blued steel and one gilt for mean time and two further gilt hands for solar time.

Lying behind the chapter ring is the year calender which rotates against a pointer above twelve o'clock and below this a little of the equation disc may just be seen. Collection Dr. Sobek. Photos courtesy Osterreichisches Museum Fur Angewandte Kunst, Wien.

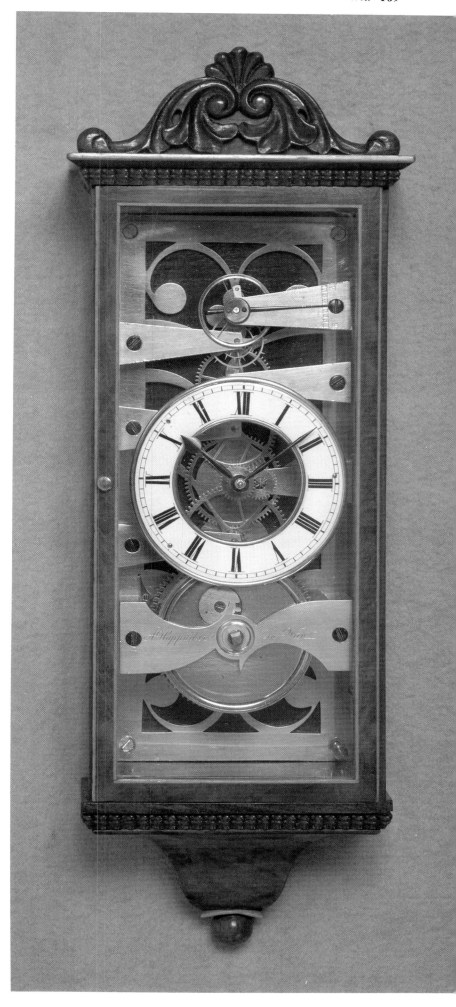

Fig. 166. A beautiful mahogany cased spring driven miniature Vienna regulator with duplex escapement, circa 1850 by Anton Happacher, Wien. Length: 11¾'' (30 cms.). Photo courtesy Museum der Stadt, Wien.

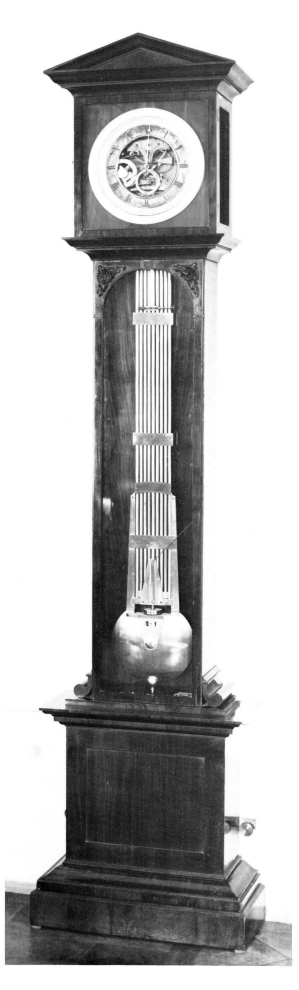

Figs. 167a, b. A complex mahogany cased skeletonized longcase regulator signed on the silvered chapter ring Binder in Wien, circa 1820-30. The year duration movement has a superbly constructed nine rod gridiron pendulum with a pointer (seen here fallen to one side) indicating the temperature, or the amount of thermal compensation being achieved. This pendulum so closely resembles those being produced in France at that time that one feels it may well have been imported. The driving weight is contained in a separate compartment at the back of the case.

The main dial has center sweep minute and hour hands and a subsidiary seconds dial above six o'clock, in the center of which may be seen the dead beat 'scape wheel which has inverted pallets. To the left is a partially concealed ring for the year calender which revolves against a pointer which may just be seen inside the main chapter ring betweem nine and ten. Behind this ring is the equation kidney which receives its drive from the great wheel via worm gearing to the wheel immediately below the date disc. A lever may be used to release this worm drive when setting the equation work. An arm held against the edge of the equation kidney is connected via a chain to the tube onto which the Solar Time hand, here seen almost upright, is mounted. Collection Dr. Sobek. **Photos courtesy Osterreichisches Museum Fur Angewandte Kunst, Wien.**

4

Belgium

HUBERT SARTON

Hubert Sarton was a man with an extremely inventive turn of mind and whereas his main preoccupation was horology his interests extended considerably beyond this field. He concentrated his efforts on the manufacture of skeleton clocks and was probably the first to do this, producing more than any other maker at this period and evolving a style which was very characteristically his own. His choice of the skeleton clock was almost certainly due to his wish to show off his ingenuity and skill as completely as possible but probably an even more important consideration was the desire of his wealthy patrons to impress their friends with their horological marvels and to gain pleasure from them by being able to look in and wonder at their innermost complexities.

To put this in context it must be realised that at that time the leading clockmakers were considered to be extremely important men and very much part of the scientific community as is evinced, for instance, by the fact that Sarton was asked by the "Prince Eveque" to start a Science Society, which rapidly acquired a prestigious reputation and became known as the "Societe d'Emulation".

Hubert Sarton was born in Liege in 1748 and from an early age showed a great aptitude to the mechanical sciences. He was apprenticed to his uncle Dieudonne Sarton, also of Liege, in 1762 and in 1768 went to Paris to work with Pierre Leroy where he undoubtedly obtained much valuable knowledge and experience which was to stand him in good stead and influence him for the rest of his life.

In 1772 at the age of 24, Sarton returned to Liege which at that time was under the rule of Austria. The representative of the House of Hapsburg from 1744-1780 was the Duke Charles Alexander, Prince of Lorraine, who was the brother of The Emperor, and it was he who appointed Sarton "Court Mechanic".

During the period 1771-84 Liege flourished under the guidance of the Austrian Prince Archbishop Francois-Charles de Velbruck known as the "Prince Eveque". Like Sarton he had spent some time in Paris where he was Minister to the Count of Outremont at the Court of Versailles and this had stimulated in him a strong interest in the Arts and Sciences. It was this which undoubtedly led to his patronage of Sarton and it would seem that their relationship was a relatively close one as he asked Sarton, as mentioned earlier, to start a "Science Society", a task which he accepted with pleasure.

Over the years Sarton made some fine clocks and watches for the Prince Eveque and undoubtedly gained important customers through him. He also made several clocks for the Prince of Lorraine who had an extensive collection. When the Prince died in 1781 a large part of his collection, 118 clocks and also 52 watches, was sold (including some pieces by Sarton) and a catalogue of these was published which is still in existence, part of which is reproduced here:

Fig. 168. Two Pillar clock signed on the enamelled dial which has an exposed center, Sarton à Paris, a signature which appears on his clocks not infrequently. It is possible that he used "Paris" instead of "Liege" because of the greater kudos this would imply or because the clocks were marketed there. It is known that on occasions he supplied clocks to some of the best Paris clockmakers as their names sometimes appear on his products.

It has center sweep seconds, minute, hour and date hands, circular plates with countwheel strike on a bell and pinwheel escapement. Height: 20" (51 cms.) Photo courtesy Sotheys, Bond Street.

A. "A pendulum clock, with equation of time, a perpetual calendar, the length of the day, the phases of the moon and the years, leap years and normal, all making its revolutions through the movement of the said piece [The pendulum ?] made by Sarton à Liege, the case is in mahogany richly decorated in gilt bronze."

B. "An astronomical clock, with 3 dials, which makes its meridian revolution in 24 hours, a small sun raises and sets at the exact time on an horizon which raises or lowers, thus marking the length of the days and nights. The clock shows also the phases of the moon and its synodic revolution; it shows with precision the hour and minute of the sideral time, and the sun's passage at the meridian, a universal calendar shows the days of the month for leap years or not. The pendulum is compensated grid-iron of nine rods alternatively of brass and iron, to.... and it follows the precepts of Ferdinand Berthoud* invented by Sarton, which was approved by the Imperial and Royal (Austrian) Academy of Sciences and Letter of Brussels in its meeting of January 24th, 1783".

However some of his finest pieces were omitted from this catalogue including a fine astronomical clock by Sarton which was subsequently acquired by Baron d'Otreppe de Bouvette and is now in the Curtius Museum in Liege. It has six dials which give the following indications:

1. The time at different places around the world.
2. Sunrise and sunset.
3. Local and siderial time.
4. Age and phases of the moon.
5. Year calendar from 1795-1844.
6. Days of the week and months of the year.

In 1783, the year before the "Prince Eveque" died Sarton was appointed City Counsellor and Treasurer. In 1794 the French Revolutionary Armies swept through Belgium and occupied Brussels and this undoubtedly had an adverse effect on Sartons business. Indeed it was probably only the small sum of money which he received from the city after his official posts were abolished at that time which enabled him to keep his business alive and feed his wife and family of 8. However it would seem likely that business improved for him in the early 1800s as a considerable number of his clocks which are still in existence, such as the multi-dialled clocks seen in Figs. 172 and 174 are believed to come from that time.

---

* What happened to Harrison? Was he not the inventor of the Gridiron?

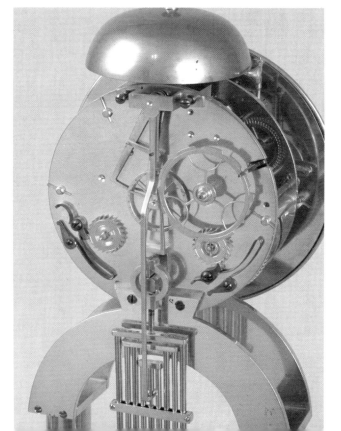

Figs. 169a, b. A similar clock to that seen in Fig. 168 but unsigned, a not uncommon occurence with Sarton's clocks. Note the external pinwheel escapement, count-wheel strike, knife edge suspension to the pendulum; beautifully arcaded spokes to the wheels and the well executed external clickwork, all signs of the fine quality of Sarton's workmanship. Height: 20" (51 cms.)

Besides his relatively simple single dialled clocks which he produced in a variety of forms such as those seen in Figs. 168-171, the multi-dialled clocks and the clock with rotating head he also produced a complex musical clock, a self-levelling table regulator which is now in the Museum der Stadt, Wien, (Fig. 178) a "Grand Pendule Astronomique" which was bought by Le Duc Charles de Lorraine, a dial operated manually, for calculating the equation of time, and various watches including a stopwatch.

One of his principal claims to fame in the field of horlogy was his self winding watch (see references 3 and 4 below) which was fitted with maintaining power and may well have been the first devised. A report on it was presented to L'Academie des Sciences de Paris by LeRoy and Fouchy in 1778.

Throughout his career he made numerous interesting inventions outside the field of horology. Amongst these were:

1. A spinning wheel for cotton and wool.
2. Machinery for extracting coal from the ground (1775).
3. A wheelchair for paraplegics.
4. A hydraulic pump which was used by the Dutch (1820-22).
5. A windmill with four horizontal contra-rotating sails.
6. An escalator which was devised for both public use and in coal mines.

Following the takeover from the French by the Dutch in 1815 Sarton's fortunes seem to have declined and although in 1822 he received a medal from the Dutch for his water pump he was given no money. He died on 28th October 1828 just short of his 80th birthday and to quote Aghib "Penniless, heartbroken and misunderstood".

It is obviously impossible to discuss this remarkable man at any greater length in a publication such as this, but for those who would like to know more about him it is suggested that they read the excellent articles listed below:

REFERENCES

1. Pholien, Florent, *L'Horlogerie et ses Artistes au Pays de liege,* Publiee Sous Le patronage De la Societe Libra D'Emulation (The Society Sarton founded) Liege, Prix Rouveroy, 1933.
2. Aghib, E., "Hubert Sarton of Liege", *Antiquarian Horology,* December, 1972.
3. Chapuis, A., E. Jaquet, *The History of the Selfwinding Watch 1770-1931,* Rolex Watch Co., Geneva, Chapter III, pages 65-68.
4. Chapius, A., E. Jaquet, *La Montre Automatique Ancienne,* Neuchatel, 1952. Addenda after page 62.
5. Defrecheux, Joseph, *Biographie Nationale des Sciences, des lettres et des Beaux-Arts de Belgique,* pages 413-419, 1911-1913.

Sarton's Single Dialled 2-Pillar Clocks.

Sarton produced what might be called his standard two pillar clocks which may be seen in Figs. 168-171. These nearly always incorporate a center sweep seconds hand which makes two revolutions a minute and have hour and minute hands of delicate design and use a gridiron pendulum. However although at first sight they all appear the same and indeed the frames and bases are virtually identical, he did vary the design appreciably; for instance some are larger than others; the dial centre was sometimes omitted; the crossing out of the wheelwork varied and whereas the majority had countwheel strike on a bell mounted above the movement, single train clocks were also produced.

One of the most interesting is that seen in Fig. 171. Although it appears like the others to have a half seconds pendulum this moves very slowly and is actually a seconds beating compound pendulum. A remontoire is also incorporated in the clock.

Fig. 170. A similar clock to that just described but with pull quarter repeat on two bells and without a cutout to the dial center. It is signed Florent Daywaile, Brussels to whom Sarton presumably supplied it. Photo courtesy Phillips, London.

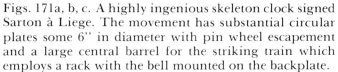

Figs. 171a, b, c. A highly ingenious skeleton clock signed Sarton à Liege. The movement has substantial circular plates some 6'' in diameter with pin wheel escapement and a large central barrel for the striking train which employs a rack with the bell mounted on the backplate.

A remontoire from the striking train rewinds the weight for the going train every time the clocks strikes, via a large wheel mounted on the backplate. A smaller wheel carries the counterweight.

A particularly interesting feature is that the 9 rod gridiron pendulum, although the same length as a half seconds pendulum, beats seconds. This is achieved by fixing to the top of the pendulum a ring of brass which is concealed behind the dial (Fig. 172c), which thus prevents us seeing any apparent reason for the mysteriously slow movement of the pendulum. Height: 20'' (51 cms.) Further views of this clock and details of the remontoire may be seen on page 135.

Sarton's 4-, 5-, and 6-dialled Clocks.

According to Pholien[1] the 4-dialled clocks were made after 1812 and they are, in many ways, the most attractive clocks he produced, with their beautifully executed enamelled dials. The layout and the information contained on these varied considerably but the overall concept was always the same.

Probably one of the first made was that shown in Fig. 172 as the movement seems to be an adaptation of his simpler movements (Figs. 168 and 169) which had circular plates. As can be seen, brackets have been screwed to these to carry the subsidiary dials. Exactly the same technique has been employed on the clock seen in Fig. 173, and indeed the whole of it, except for the treatment of the dials and the hands, is typical of Sartons work. Interestingly it is signed by N.M. Lhoest his brother-in-law and it would seem likely that Sarton supplied the basic clock and Lhoest fitted the dials and hands.

It is probable that the clock with a World Time dial at the top (Fig. 175) was made a little later than those seen in Figs. 172 and 173 as in this example Sarton has tidied up his design, eliminating the rather crude brackets carrying the subsidiary dials by adopting much larger basically triangular plates for the movement.

The 5- and 6-dialled clocks are undoubtedly two of the finest clocks ever produced by Sarton and are different from most of his others in that he uses scroll or arched supports for the movement rather than the two pillars.

---

[1] See references on page 173.

Figs. 172a, b, c. Four-dialled clock by Sarton circa 1812-1815. It has a pin wheel escapement, an attractively crossed out countwheel for the bell strike, beat regulation to the gridiron pendulum which rests on a knife edge and centre sweep seconds (missing when photo taken), minute, hour and date hands. It shows moon phases in the arch, days of the week with their relevant diety to the bottom right and months with their zodiacal signs to the bottom left. Points to note are the apparent adaptation of Sartons original circular plated movement with brackets to take the calendarwork and the use of bolts to hold the two halves of the brass case of the pendulum bob together, a characteristic feature of most of Sartons Pendulums. Collection Klokkenmuseum. Schoonhoven, Holland.

173b.

173a.

Figs. 173a, b. A clock very similar to that seen in Fig. 172 and almost certainly made by Sarton but signed by N.M. Lhoest his brother-in-law who was probably responsible for the dials, hands, and bezels as none of these is characteristic of Sartons work. The Roman numerals are much heavier; the bezels, which are beautifully engine turned, are much thicker than those favoured by Sarton and are more like the work of Verneuil, and other French makers of that period and the hands lack the delicacy of those normally employed by Sarton. Collection Klokken-museum. Schoonhoven, Holland.

Figs. 174a, b, c, d. *5-Dialled Clock.*
Although one cannot be certain that this clock, which is unsigned, is by Sarton, so many of the features are characteristic of his work that the probability is that he made it, particularly as it is one of a group of clocks known to be by him which are at the Museum at Schoonhoven, Holland.

The movement plates are basically triangular in shape with a cut out at the top for the sunrise/sunset mechanism. It has countwheel strike (bell missing) with a vertical arbor to the hammer and a fine gridiron pendulum (Fig. 174d.) which indicates the degree of thermal compensation. Dead beat escapement is used with adjustable pallets and beat regulation.

The main dial consists of two parts; an outer chapter for seconds, minutes, hours and year calendar with every day marked on it and also the seasons and the zodiaccal signs, and in the centre is shown the moons phases. To the bottom left are the days of the week and on the right a double 12 hour dial gives the time of High Water. The World Time dial is similar to that seen in Fig. 175b. Photos taken by Courtesy of the Klokkenmuseum, Schoonhoven.

Figs. 175a, b, c. *A Complex World Time Clock by Sarton.*
This clock would probably have been made a little after
those shown in Figs. 172 and 174 as large triangular plates
have been adopted, thus eliminating the need for brackets
to be screwed to them to carry the subsidiary dials.
However the other mechanical details are similar to the
clocks already described.

The main dial has very delicate centre sweep seconds,
minute and hour hands and subsidiary dials for days of the
week, days of the month and months of the year. To the
bottom left is shown the phases and age of the moon and
on the right the times of sunrise and sunset and thus the
length of night and day. At the top is a dial which
continuously indicates the time at 52 places around the
world by means of a rotating outer enamelled ring. There
is a similar clock in the Haussner Restaurant, Baltimore,
U.S.A. Photos taken by Courtesy of Sothebys, Bond Street.

Fig. 176. This clock is very different from any of those described so far. The skeletonised movement is roughly triangular in shape and has Dutch strike on two bells. It employs a pin wheel escapement and a brass bob pendulum with silk suspension which is adjusted by a hand at the top of the dial. This is made of glass, being painted white on the back and having the numerals and subsidiary dials etched out and painted black.

The main dial has a centre sweep seconds hand which goes round twice a minute and a centre date. The subsidiary dials at the bottom are for days, months and age and phases of the moon. It is signed H. Sarton, Clockmaker & Mechanic to Their Highnesses at Liege'' which Aghib[2] suggests may refer to the Duke of Saxe-teschen and his illustrious wife. Height: 1' 5¼'' (44 cms.)
Photo courtesy Christies, London.

177a.

177b.

Fig. 177a, b, c, d, e. *Skeleton Clock with Swinging Dial made by Sarton for the Duke of Lorraine 1778.*
Although this clock is unsigned, possibly at the request of Duc Charles de Lorraine, there can be no doubt that it was made by Sarton as its' description exactly matches that contained in Sarton's book, (Fig. d) a copy of which is in the possession of the British Horological Institute. It is also referred to (Fig. e) in the catalogue of the Duke's clocks and watches when they were disposed of shortly after his death.

The main dial which is enamelled and has an ormolu bezel has the possibly unique feature of swinging between 3 set positions, left; centre and right, moving once each minute, presumably so that the time can be seen clearly from all parts of the room in which it is situated. It can also, if required, be locked in any one position. The main dial which has centre sweep minute, hour and date hands has a relatively heavy counterbalance and its speed of movement is governed by a fly.

A massive mainspring powers the movement and this now gives a duration of approximately 5½ days; however, it is likely that the spring has been shortened over the years and that originally the clock was designed to go for at least a week. The mainspring activates a remontoire which lifts the going weight a little over 1¾" ever time the dial moves i.e. once a minute.

A second fixed dial immediately above the main dial indicates days and months and has a centre sweep seconds hand.

The substantial frames of the movement are held together by 4 pillars fastened by knurled nuts and have arms extending from either side, one to carry the pendulum and the other the remontoire weight. Graham dead beat escapement is used and the pallets of this are linked by a relatively complex series of rods. The five rod gridiron pendulum has knife edge suspension together with screw regulation at the top so that it may be adjusted without stopping the clock.

Five spoke wheelwork is employed and there is count wheel quarter striking on two bells mounted on top of the clock, one inside the other. The massive winding key some 8" long may be seen in the rear view of the clock. Photos taken by courtesy of Mr. A. Odmark.

177c.

DESCRIPTION

ABRÉGÉE

DE PLUSIEURS PIECES

D'HORLOGERIE,

Qui sont approuvées de plusieurs Académies, et ont été inventées ou exécutées par HUBERT SARTON, Horloger-Mécanicien, demeurant à Liege, à l'entrée du Pont-d'Isle, et pendant la Saison des Eaux, à Spa.

On y a joint deux Projets ; l'un, d'une nouvelle Machine à extraire la Houille ; l'autre, d'une reconstruction de la fameuse Machine de Marly.

—————

AVIS.

ON trouve à Liege et à Spa, chez HUBERT SARTON, un assortiment des plus complets en toute espece d'Horlogerie, dans le goût le plus nouveau, comme montres d'or et d'argent de tout genre, uniës ou émaillées,

A

177d.

(17)

III.

PENDULE DE COMPAGNIE.

Cette Pendule, qui obtint l'approbation de l'Académie Royale des Sciences de Paris, en 1778, comme étant bien exécutée et bien construite pour produire ses effets, fit aussi partie des précieux meubles du cabinet du feu Duc Charles de Lorraine, etc. La pareille se trouve encore actuellement chez l'auteur. Elle a la propriété de montrer l'heure et la minute précises sous plusieurs points de vue et dans différentes places d'un appartement à la fois, en promenant son Cadran horizontalement sur une ligne circulaire et en la décrivant, en trois mouvemens égaux par moitié de circonférence dans l'espace d'une Minute. Le Cadran s'arrête même pendant quelques Secondes de distance en distance, pour donner la facilité de reconnoître les points indiqués, et ayant décrit le demi-cercle il revient sur la même direction en rétrogradant à pareils intervalles. On peut accélérer ou ralentir sa marche à volonté et l'arrêter par le moyen d'un Poussoir sans rien changer au Mouvement de la Pendule qui va huit jours sans la remonter.

C

177e.

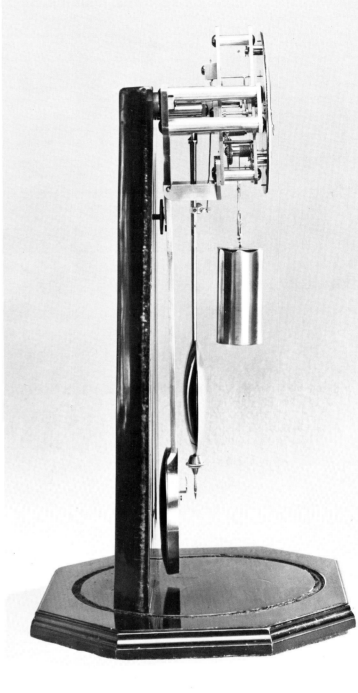

Figs. 178a, b. *Sarton's "self-levelling" table regulator.*
It would seem likely that this regulator, which is designed to be self levelling, was probably made for field use, possibly in connection with surveying or astronomical observations. It is now in the "Museen der Stadt Wien" who kindly supplied the photographs and it is tempting to think that it may have found its' way to Austria at the time when Belgium was still being ruled by them.

The whole movement is pivotted along both axes like a boxed chronometer and is kept vertical by a heavy counterweight which may be seen in the side view hanging down behind the pendulum. It can compensate for a tilt of up to 6 degrees left to right and 3 degrees front to back.

It employs a pinwheel escapement and a gridiron pendulum which is provided with beat regulation. The dial has a centre sweep seconds hand which goes round twice a minute even though the outer dial is graduated 0-60. Minutes are shown at the top of the dial and hours below. Height: 21" (53 cms.) Photos courtesy Museum der Stadt, Wien.

Figs. 179a, b. *Skeleton Clock signed M.J. Baty. Fils a Liege.* Whilst this clock bears a superficial resemblance to Sarton's 4-dialled clocks the workmanship and style are very different. It has a pinwheel escapement mounted on the backplate with an inverted crutch rising up to the pendulum rod which is fitted with beat regulation. The bob is formed by a circle of brass beads surrounding the main dial.

The top dial shows the age and phases of the moon, the main dial has centre sweep second, minute and hour hands; to the bottom left are shown the days of the month and the months of the year, and the dial on the bottom right has a double hand which indicates the days of the week on one end and the ruling diety on the other.

180a.

5

Holland

Despite a diligent search which included contacting the leading museums and dealers in Holland and also some collectors it only proved possible to find three skeleton clocks which were definitely of Dutch origin which is somewhat surprising. Whilst it is realised that some may well have been overlooked it does seem that only a very small number of skeleton clocks were made here.

180b.

180c.

Figs. 180a, b, c. This relatively complex clock has a somewhat crudely engraved dial, believed to be of pewter, with two inner dials, one calibrated in Roman and the other in Arabic numerals which show both Mean Time and Local Time.

The anchor escapement regulates the movement of the right hand dial. There is a seconds hand below 12 o'clock with the 'scape wheel on the same arbor. Also fitted to this arbor is a crown wheel, the teeth of which rotate a shaft by engaging a tooth on it once every second. This drive controls the movement of the left hand dial and the centre sweep seconds hand.

Both trains use the same mainspring, that on the left, which has an extra wheel, providing the power for the count wheel strike. The pendulum, which is characteristic of van Spanjes work is an interesting variant of Lepautes compensated pendulum with a subsidiary dial on it calibrated 0-20 twice and marked Koud and Warm. The clock is signed C. van Spanje, Tiel No. 15 and would date circa 1840. Photos 180a, b. kindly supplied by Stender B.V., Holland, and Photo 180c taken by courtesy of The Klokkenmuseum, Schoonhoven, Holland who also supplied the description.

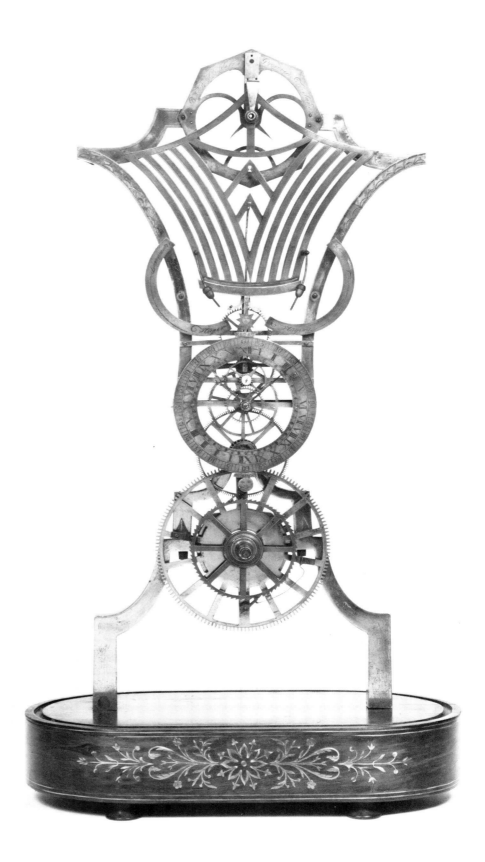

Fig. 181. A most unusual long duration Dutch skeleton clock with a gilded frame in the form of a tulip, the vertically planted train with high count wheels and pinions; a large 14 spoke great wheel with visible ratchet wheel with double clicks and dead beat escapement. There is an engraved "sector" pendulum with knife edge suspension resting on a pierced support spanning the top of the frame which is signed B.P. Gilder. Provencie, Friesland, with regulation arcs on which are engraved Langzamer & Huger. The double 12 hour chapter ring has Medag & Mednaacht marked upon it. Height: 2' 5'' (73.5 cms.) Photo courtesy Christies, King Street.

Fig. 182. A small timepiece skeleton clock signed on the enamelled dial by F. C. de Jong, Amsterdam (a well known maker of electric clocks) and dated 1836. It is believed to have been his apprentice piece. It has anchor escapement with silk suspension to the anchor shaped pendulum. Height: clock only—7¾'' (19.7 cms.) Photo taken courtesy of the curators Klokkenmuseum, Schoonhoven, Holland.

6

America

The first domestic clocks to appear in America would almost certainly have been lantern clocks which were taken over from England by the early settlers. As the 17th century progressed and the longcase clock came into being it is likely that some of these went out to America also although the numbers would have been relatively few as only a small number of longcase clocks were being produced in the United Kingdom at that time and both their size; the difficulty of transporting them and their cost would have been prohibitive for the vast majority of immigrants.

Amongst the early settlers were a steadily increasing number of black-smiths and clockmakers and at a relatively early stage in the birth of America they would have started producing clocks. Boston for instance was an early centre of clockmaking and had by 1688 produced a Turret clock.

Abel Cottey who came from Crediton in Devon was probably the first authenticated clockmaker in America. He was born in 1655, emigrated to Philadelphia in 1682 where he is recorded as having made longcase clocks and died in 1711. His business was obviously a prosperous one as he acquired much real estate.

By the beginning of the 18th century clockmaking was fairly firmly established in New York; New England, Pennsylvania and Virginia and followed very much in the styles prevailing in England at that time. A system of apprenticeships was established similar to that which already existed in Europe.

As the century progressed clockmaking gradually became established over a much wider area although certain places such as Connecticut became centres for the craft as did for instance Liverpool and Bristol in England and, at an earlier date Augsburg and Nurenburg in Germany.

It is likely that the methods of clockmaking in the major centres were very similar to those in England until around 1780-90, although local woods would generally have been used for the cases; however in the remoter areas, because of the lack of facilities and in particular materials such as brass, considerable ingenuity had to be exercised. Thus brass was used in strip form rather than as plates and wooden movements were sometimes constructed.

Although the import of complete clocks from the United Kingdom continued throughout the 18th century it is likely that considerable numbers were also supplied in the form of either movements; dials or just components, as occurred in the country districts of England, and that the majority of cases were made locally.

Virtually no watches were made in America during the 17th and 18th centuries, it presumably being far cheaper to import them from Europe. The first recorded watchmaker in America is Thomas Harland who emigrated from England in 1773 but it is unlikely that he made many in the U.S.A.

Similarly very few bracket clocks seem to have been made up until the War of Independence. This was probably largely due in part to the difficulty of obtaining suitable springs other than by importing them, as there were no rolling mills capable of producing good spring steel in America at that time, and also the greater complexity of these movements.

Figs. 183a, b. This clock, photographed prior to restoration, is very similar to that illustrated and described in Figs. 185a, b, c. However it is somewhat later as it is signed in the centre of the lower dial "Boston Clock Company". A.D. Cranes Patent 375 days Patented Feb. 10th 1841. Extended Dec. 15th 1851. Improved June 22nd 1852 and Jan. 9th 1855.

The movement is simplified in that there is no tidal indication and the under-dial wheelwork is solid. The year calendar is not visible in this picture but the shutters for sunrise/sunset may be seen and these would appear to operate auto-matically, the wheelwork for this being shown in the rear view. Photographed by courtesy of Mr. J. Fanelli, New York.

Although some very fine longcase clocks were made in America during the 18th century it is likely that they were mostly made to order. However all this was to change following the War. After this there was a shortage of materials and a need for cheaper and smaller clocks in larger numbers and thus local designs were developed.

Willard in Grafton Massachusetts was one of the first to evolve his own style of case and movement. He produced his celebrated Banjo Clock with a metal movement and by 1840 had made 4,000 of them.

Machinery was now starting to be used for clockmaking and it was evolving quite rapidly from a craft to a manufacturing process. Gideon Roberts (1749-1813) of Bristol, Connecticut, probably gaining much of his knowledge from immigrant Dutch and German clockmakers, started to produce wall clocks with wooden movements in relatively large numbers and established an assembly plant in Richmond, Virginia around 1800.

However it is Eli Terry who is credited with making the first mass produced clocks when he accepted and fulfilled an order for some 4,000 in 1806. From there he rapidly developed his sales and manufacturing techniques and other makers such as Seth Thomas and Chauncy Jerome joined in, the latter using brass strips to make his clocks. Such was their energy and ingenuity that by 1845 production was rising towards a million a year and they were looking to Europe and other export markets to absorb their production.

The early mass produced clocks were mostly timepieces only i.e. they did not strike but as time went on this gradually changed. Similarly whereas virtually all the early clocks were weight driven by around 1845 springs started to be used extensively and shortly afterwards the balance wheel was adopted on some clocks so as to make them portable.

It is widely assumed that by 1825 mass production had virtually destroyed the traditional craft of clockmaking in America but this is by no means true. Some exceptionally fine clocks, particularly in the field of precision timekeeping, were made in America during the 19th century as anyone who has visited the Time Museum in Rockford and various other American museums and collections will realise.

Unfortunately relatively few skeleton clocks were made in America but even some of these such as Cranes torsion clocks, (Figs. 183, 184, & 185) Marshs' fine double dialled skeletonised regulator with grasshopper type escapement (Fig. 189), the anonymous timepiece with helical gearing (Fig. 190), and Pardees' weight driven clock (Fig. 188) give some idea of the ingenuity of the designs evolved.

The only two which seem to have been produced in any numbers are the box calendar skeleton clock made by Ithaca (Fig. 193) and those of Silas B. Terry (Figs. 191 & 192). However the total number made of both these clocks, although not known with any accuracy, must be negligible by comparison with other clocks which were mass produced in America in the 19th century. The Silas B. Terry clocks were interesting in that they were timepieces but usually employed twin going barrels and a large vertically mounted balance wheel at the top of the movement which had a centre sweep seconds hand. It may well be that the skeleton clcok was chosen to show off the balance wheel escapement which had only come into general use in America a relatively short while before.

AARON D. CRANE

Aaron D. Crane was born on May 8th 1804 and from an early age showed a keen interest in clocks: However he served no apprenticeship and appears to have been self taught, working throughout his life in isolation from other clockmakers.

In March 1829 he was granted a patent for "An improvement in the plan of constructing clocks" which described a clock of radical design incorporating many new concepts for the escapement, pendulum and motionwork. Unfortunately the details of this were lost in a fire in the Patents Office in 1835.

In 1830 he patented a turret clock design and in 1835 is listed as a clockmaker at 15 Lawrence St. Newark and by 1840 had moved to 6 Halsey Street. In that year he met Abraham V. Spear, a lawyer; Banker and Entrepreneur who backed him to make a year clock. A patent no. 1973 was issued on 10th December 1840 for a single ball torsion pendulum year clock and in the following year he gained a silver medal at the October American Institute Fair at Niblo Garden. New York City.

To appreciate the size of Cranes achievements it must be remembered that nearly all the ideas and mechanical features which had appeared in American Horology up until this time had originated in England, France and Germany. Aaron Dodd Crane had invented an entirely new type of clock known as the "Torsion Clock", which was destined to be produced in Europe in very large numbers[1] in the second half of the 19th century and throughout the 20th century.

---

[1] Shelly, F. "Aaron Dodd Crane, An American Original," N.A.W.C.C.
Inc. Columbia, Pennsylvania. Bulletin number 16. Summer 1987.

Fig. 184. A striking weight driven skeleton clock by Aaron D. Crane in the New Jersey State Museum Collection Trenton which was a gift from James R. Seibert. It was made for the Fifteenth Annual Fair of the American Institute in New York City which was held 12th October, 1842[2].

The overall concept is very similar to that of some of the skeleton clocks being produced in Austria somewhat earlier in the century in that it is weight driven and the movement is supported by twin pillars. The chapter ring is delicately fretted out as also are the plates. The lines from the weights run over pulleys above the movement and then down to the barrels so as to give the maximum fall. Cranes torsion pendulum is employed with three brass balls mounted on the base.

---

[2] Drost. *Clocks & Watches of New Jersey.* P. 86.

185a.

185b.

Figs. 185a, b, c. Year duration spring driven *Torsion Pendulum Timepiece with Astronomical indications* signed on a plaque above the top dial "Made and Patented by A. D. CRANE, NEWARK, NEW JERSEY, 375 DAYS."

This timepiece sits on a double marble base from which rise 4 tapering cast brass pillars each of which is fabricated in two parts. These support a brass box containing the astronomical work and in the centre of which is the dial for this.

Rising from this box are 4 brass scrolls which carry the main movement and dial. The whole is held together by threaded iron rods running from the bottom of the scroll pieces to the base. The wheelwork is 5 spoke and lantern pinions are employed.

Aaron Crane's "Walking Pawl" escapement

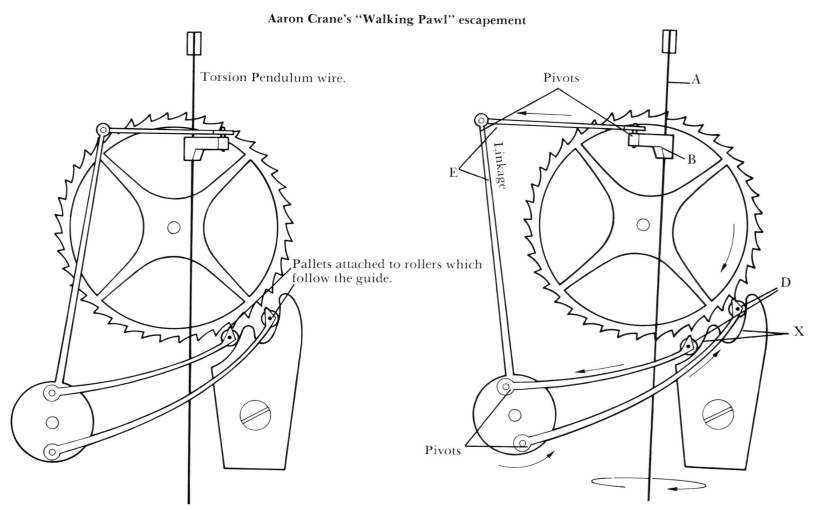

Fig. 185c. **Aaron Crane's "Walking Pawl" escapement**
1. Torsion pendulum at rest, about to turn clockwise.

2. In turning clockwise, block (B) attached to wire (A) turns thus imparting, via linkage (E), the change of pallets. This escapement has recoil. Drawing by D. Penney.

The indications given by the Astronomical work are:
a) *Year Calendar* with the signs of the Zodiac on a silver ring approximately the same diameter as the opening in the bezel, but set back 1" behind it and pivotting on two date ring rollers.
b) *Times of Sunrise and Sunset.* However this is not an automatic function, as the bell crank which would alter the length of the time of day, which is obviously continually changing, has to be adjusted by hand and thus it is a very impractical design. Moreover the method of adjustment which involves using a small screwdriver inserted from underneath the pillars requires considerable knowledge and skill and to maintain any degree of accuracy would have to be performed at least every two weeks.
c) *The Passage of the Sun* through the heavens during the day.
d) *The Passage of the Moon,* which is represented by a solid ball some 1½" in diameter. The moon which is painted black on one side and gold on the other also rotates to indicate its' phases.
e) *A Universal Tidal* dial which may be adjusted to indicate the times of high water at any port.

An unusual feature on this clock and also the other similar example of Crane (Fig. 183a) is the convex profile of the fuzee. This is because the gut line, which is attached to the large end of the fuzee, is wound onto the train barrel situated on the right side of the timepiece. The fuzee thus works in the opposite direction to a conventional one, the spring being fully wound when the gut is pulling on the equivalent diameter of the fuzee and vice versa.

The torsion spring is suspended from a replacement brass cock which is screwed to the inside of the front plate. Just below it is a rack pinion which regulates the timepiece by raising and lowering a pair of chops which embrace the spring and effectively lengthen or shorten it. A regulating hand with slow/fast indication may be seen inside the chapter ring just below 12 o'clock.

The torsion pendulum some 1½ lbs in weight is in effect a carousel with 8 separate decorative weights suspended from a frame above them. On the base is a cast brass figure of a seated girl holding a basket of fruit.

The dial is of reverse painted glass with a white background. There is a cut out between II and III for the winding key and the hands are counterpoised. Photos courtesy National Museum of American History who kindly also supplied a very good description.

Its final development might be considered to be in the Atmos clock for which it is especially suited as only minimal power is applied to the train by any change in temperature or atmospheric pressure.

The principle of the Torsion Pendulum is similar to the balance and hair spring but makes use of a bob, which rotates back and forth and, as it does so, twists and untwists the thin steel strip by which it is suspended. At the same time the escapement is released and impulse is then given to the bob via the suspension. (Fig. c)

The big advantage of this system is the very slow rate of vibration and the small amount of power required which enables a long duration, commonly 1 year, to be achieved with a relatively small mainspring.

From 1842-1848 Crane became known as the "One year clockmaker and J R Mills and Co manufactured his clocks. During this period 5 different versions of the year clock were made. Month duration clocks were also constructed and in 1845 mention is made of a four year clock. Eight day clocks were introduced in 1846 and in the same year the company was reorganised and called The Year clock company". Sadly in 1848 the company ceased trading with substantial losses due, at least in part, to the economic decline at that time.

In 1849 Crane set up business on his own account as the "One year Clockmaker" and worked at 6 Lombardy Street, Newark until 1856. During this period many of his clocks were sold under the Boston Clock company label.

In 1850 he produced his first Astronomical timepiece which can only be described as a "tour de force". In all 5 of these clocks were made of which two are illustrated here.

The "Walking Pawl" concept for the escapement of his clock was refined and subsequently applied to such diverse uses as fans for ventilation; a self rocking baby cradle and a revolving bookcase.

In 1854 he registered a patent for multi-task automated machinery which could be applied, amongst other things, to complex turning and in 1855 another patent was registered which related to temperature compensation for the torsion pendulums of his 6 ball clocks. He died some 6 years later in 1860.

Throughout his life he turned his mind to the invention of many different types of machinery but always seems to have kept returning to the perfection of his year clock to which so much of his life was devoted.

Few records seem to exist of the clocks which he produced or were made to his designs during his lifetime: However Frederick Shelley in his excellent and carefully researched account of Aaron Dodd Cranes life and work records that at the present time the following clocks are known to still exist.

1 Ball year clocks—4	CT movement 8 day clock—1
2 Ball month clocks—2	Astronomical timepieces—4
3 Ball month clocks—13	Tower clocks—3
3 Ball 8 day clocks—13	Clock patent models—2
6 Ball year clocks—31	Total—73

WAGON-SPRING CLOCKS.

This chapter would not be complete without mention of that fascinating clock the Wagon-spring which is peculiar to American clockmaking. In this a leaf spring similar to that used on wagons, coaches and indeed many cars is employed to supply the power. Technically the problem with using a spring such as this is the springs strength and its limited range of movement.

It is interesting that as early as 1680 Lagvile working at Chaalons in France made a clock driven by a powerful leaf spring[8] which is now in the Chateau des Monts, Switzerland. However this would appear to be an isolated incident and the only other use of a similar spring by clockmakers is confined to those seen on the pull quarter repeat work of early English bracket clocks.

It is postulated that the use of the Wagon Spring in America arose largely because good quality spring steel could not be produced locally. Indeed spring driven movements do not appear in American Clocks in large numbers until around 1840.

Joseph Ives an ingenious maker, was born in Bristol in 1782 and started business on his own account in 1812. He invented the roller pinion and went on to invent the wagon spring clock around 1825 when he was working in Brooklyn, New York. Various different types are shown in a *Treasury of American Clocks* pages 95-102.

The basic principle of the Wagon Spring Clock, illustrated in Fig. 186a is the use of a laminated or occasionally a single leaf spring, which is fixed at the centre and then tensioned by drawing up cords at either end which are fixed to the winding barrels. They are usually either of 1-, 8- or 30-days duration, and are contained in shelf clocks.

Joseph Ives produced his Wagon Spring Clocks in Brooklyn from around 1825 to 1830. At that time he got into financial difficulties and joined John Burge in Bristol where some clocks appeared in their joint names around 1832. In 1844-48 his Wagon spring clocks were made with steeple on steeple cases by Birge & Fuller and from around 1850-56 Irenus Atkins was making them with shelf and wall clock cases.

There is considerable controversy as to whether any Wagon Spring skeleton clocks were ever made, some authorities believing that those which do exist have been made up from movements which were originally in wall clocks. As an outsider I would hesitate to enter this controversy, particularly as the existence of such clocks enables me to include such an interesting item in the book. Moreover it is an ideal clock to appear in skeleton form, displaying as it does the unconventional power source so clearly.

Shown in Figs. 186a, b. is one from the Time Museums fine collection which is signed by Irenus Atkins, Bristol, Connecticut. As can be seen from the detail it has Ives' other favourite feature, rolling pinions.

Figs. 186a, b. 8-day skeleton clock powered by Joseph Ives patented wagon spring and made by Irenus Atkins, Bristol, Connecticut, circa 1850-55. It has 6-spoke wheelwork, steel strip pallets and Ives rolling pinions. Height: including glass dome and base-16''. Collection Time Museum, Rockford, U.S.A.

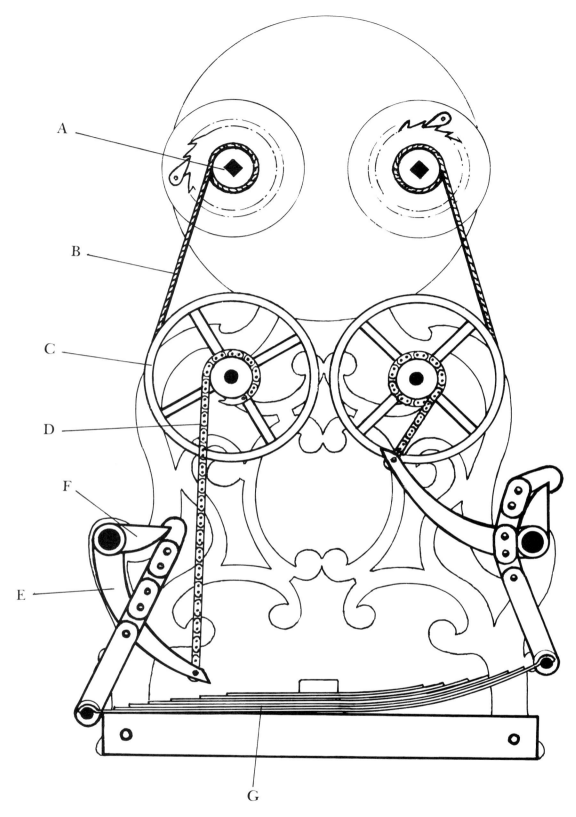

Fig. 187. *American Wagon Spring Clock*
The clock is illustrated with the left hand train unwound
and that on the right wound. Applying the key to the
winding square A, the line B is drawn up and rotates the
barrel C thereby wrapping up the light chain D. This, in
turn, draws up the lever E which has an integral tail F.
Hooked over F is a heavy link chain which connects to the
outer end of the "wagon spring" G. This spring is firmly
fixed at its center point and, when fully wound, deflects as
seen at its right hand end. Drawing by J. Martin.

OTHER MAKERS

Figs. 188a, b. An interesting weight driven skeleton timepiece with a lyre shaped decorative front plate behind which are two steel strips which carry the weight driven movement. The rear view shows the two wheel train with the unusual feature of the line winding directly onto a small barrel situated around the great wheel arbor.

As a winding square could obviously not be provided in front of the barrel a separate line is taken from it to a second barrel and arbor, complete with front winding square, situated immediately below the dial.

The recoil escapement has a 60 tooth 'scape wheel with the pallets only spanning 5 teeth. Silk suspension is provided for the half seconds pendulum which has a wire rod and brass bob.

It is signed below the base, which is decorated with a classical scene PARDEES PATENT. This type of clock is

referred to by Brooke Palmer in his *Treasury of American Clocks* on pp. 47-50, in which he lists on p. 356 William Pardee of Auburn, New York as having patented a Lyre like timepiece with ladder movement (Ray McKinney Collection) and in his *Book of American Clocks* page 254 he lists him as follows: "Pardee, William Albany, N.Y. Dir 1834-35. Patents Albany May 22, 1835. "Clocks and Timepieces" Poughkeepsie, N.Y., Feb. 10, 1836. Timepieces." So it would seem likely that William Pardee was the maker of this clock.

The only other Pardee recorded is Enoch who is listed in Carl W. Dreppard's book *American Clocks and Clockmakers* and was also working in Poughkeepsie, N.Y. but around 1840-45 so it is probable that he was a relative. Bryson Moore Collection.

189a. Front and back.

Figs. 189a, b, c. A fine skeletonized double dialled regulator resting on a white marble plinth, made by Oliver Marsh of Newark, N.J. circa 1870. It stands on a 4 ft. high marble pedestal through which the weight descends.

One dial has centre sweep minute and hour hands which are very finely executed and a subsidiary seconds ring. The second dial is of English Regulator layout with a centre sweep minute hand and subsidiary rings for seconds and hours. It also has a centre sweep date hand which reads from within out the month of the year; the day of the month and the day of the year calibrated 1-365. It has twin driving weights wound through the calendar dial and maintaining power is provided to both barrels. All the wheelwork, particularly the 'scape wheel, is very finely executed and the engraving is of excellent quality. The escapement which employs two sets of pallets is similar to a grasshopper, and a mercury compensated pendulum is used. Figs. 189b, c. Courtesy Time Museum Rockford, U.S.A.

Figs. 190a, b. A highly individual unsigned skeleton clock from the collection of Dr. S. P. Lehv believed to have been made in America, circa 1840-50. It has a going barrel with a three wheel train which employs helical gearing throughout with the wheels meshing directly with the centre and 'scape wheel arbors.

The Woolfs Tooth 'scape wheel has 120 teeth with a narrow anchor spanning only 7 of these. The chapter ring is similar to those employed on English skeleton clocks at the same period but the frame, although also basically similar, is lighter in construction and the pillars do not correspond to any of the designs seen on English skeleton clocks. The pendulum has a thin metal rod with brass bob, silk suspension and front regulation. Height: 11''. Photos courtesy Arnold F. T. Kotis.

Fig. 191.

Figs. 191 & 192. Two Silas B. Terry (1807-1876) 8-day skeleton clocks with Eli Terry's patented (1845) vertically mounted balance wheel escapement and centre sweep seconds hand (that on one of the clocks being missing). One (Fig. 191) has a single going barrel and although unsigned would almost certainly be by Terry, is still in its' glazed wooden protective case and the other which is signed on the frame Silas B. Terry has two going barrels and appears to lack any cover. Both clocks may be dated circa 1855. Fig. 191 courtesy Yale University Art Gallery. Fig. 192 courtesy Arnold F.T. Kotis.

Fig. 192.

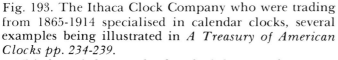

Fig. 193. The Ithaca Clock Company who were trading from 1865-1914 specialised in calendar clocks, several examples being illustrated in *A Treasury of American Clocks pp. 234-239*.

This box skeleton calendar clock has a walnut case; nickel plated bell, pendulum, movement and calendar and brass dials and frame. The 8-day two train spring driven movement has a calendar dial with a centre sweep hand indicating the days of the month and apertures inside the chapter ring giving the days of the week and the months of the year.

The movement was made by Samuel E. Root of Bristol, Connecticut (1820-96) a nephew of Chauncey Ives. The patent date on the clock is 13th July 1869 and these patents were registered to H. B. Horton. Photo courtesy Joe J. Brincat.

Figs. 194a, b. An eight-day skeleton timepiece signed along the bottom of the frame "B.C. Phelps, Wethersfield, Ct. 1865".

B.C. Phelps (1810-1896) was a Methodist Minister who seems to have specialized in making skeleton clocks as at least five examples of his work are known to exist. Whether he was an amateur or professional clockmaker is not known but the former would seem the more likely as his first recorded clock was made in 1865 when he was 55 years old and his last which had strikework was made in 1891 when he had passed 80. Photos courtesy James W. Gibbs.

Fig. 195. An unusual unsigned fuzee skeleton timepiece only 3″ tall with heart shaped frame surmounted by a platform escapement with a solid horizontal balance. Photo courtesy Arnold F. T. Kotis.

Figs. 196a, b. A small skeletonized clock known as the "Batsford Patent Timepiece" which was made by the Jerome Manufacturing Company for the Paris exhibition of 1852. Similar clocks were made with a Papier Mache instead of a brass front. Bryson Moore Collection.

197b.

THE COSMOCHRONOTROPE

This fascinating clock, if such it may be termed, which now resides at the Smithsonian Institution in Washington was conceived by James F. Sarratt a jeweller and optician of Streubenville, Ohio USA and is described in considerable detail in US patent no 220036 issued on 30th September 1879. The original concept was to sell clocks of this design as a teaching aid to those studying astronomy but the cost of its construction together with various other technical problems meant that it was never a commercial success and only the one was produced.

The clock was made to Sarratt's design with some relatively minor modifications, by **P.G. Giroud** of New York City in 1880-81 and was shown at the Columbian exhibition in Chicago in 1893. The name Cosmochronotrope derives from Cosmo (Astronomical); Chrono (Time) and Trope (turning or rotation).

The indications which it gives are:

Mean time
Siderial time
Right ascension of the sun
Position of the sun relative to the earth
Date and position of the sun in the Zodiac
Portions of earth in sunlight and darkness
Local time of sunrise and sunset for any location and date
Local sun time for any meridian and thus the equation of time.

A layout of the gear trains of the clock as published in the N.A.W.C.C. *Bulletin* of Feb. 1988 (Fig. 197e) gives an excellent idea of the overall concept of the design with the Tourbillion cage carrying the lever escapement to the bottom right and the going barrel on the left.

The finishing is excellent, the whole being machined on all surfaces and having damascened patterns on the spokes. Bevel gears are produced where necessary to allow for the 23½ degree tilt of the earths rotational axis.

The clock is of 8 days duration, mean time being displayed on the main dial shown in Fig. 197b with seconds being shown on the dial above it. However, because of the gearing used to drive the tourbillon, this hand rotates anti-clockwise. As originally conceived by Sarratt the earth globe was designed to rotate in mean time: However, when the clock was

197a.

overhauled in 1952 by Dr. Arthur Rawlings, it was converted to sidarial time. A clutch fitted to the driving gear may be used to disengage the globe, sun and auxilliary dials from the time train for demonstration purposes.

Two pairs of dials are mounted on the rear of the clock (Fig. 197c) those on the left giving siderial time, whilst those on the right show the right ascension of the sun hours, minutes and seconds.

The vertical shaft at the rear of the globe (Fig. 197c) drives two large ring gears. The upper one carries the sun which of course rotates around the earth once per year. The central shaft of the sun ring gear also supports two vertical sectors which surround the earth and divide it into daytime and night-time and can be used to measure the local times of sunrise and sunset anywhere.

Because the Earth's orbit around the sun is eliptical, its velocity varies during the yearly cycle. It is this variation which gives rise to the difference in solar and Mean solar (our) time which is called the "Equation of Time". Most ingeniously the sun ring gear is cut with 230 teeth in the half circle closest to the sun, and 237 in the other half. A stationary ring around the sun ring gives the yearly calendar and the positions of the sun in the Zodiac.

The shaft that drives the sun ring also drives a ring gear which rotates around the equator (Figs. 197b, c). This is divided into 24 hours and subdivided into minutes. This wheel also has an unequal number of teeth in its 180 degree sections, there being 260 from VI to VI on one half, and 268 on the other. By taking a meridian from any other location on the globe to the equatorial ring gear, the local solar time can be read, and from this an approximation of the equation of time obtained.

The clock was at one time fitted with an external cage representing the celestial sphere, the rotational axis of which aligned with that of the Earth and was driven in contra-rotation to the sun. To this could have been fixed distant stars, including what would have been known at that time, as time stars because of their great distance away and thus apparent invariability.

For further details of this fascinating clock, the reader is referred to R.S. Edwards' excellent article in the February 1988 issue of the NAWC Bulletin entitled *"The Cosmochronotrope, An Astronomical Clock at the Smithsonian."* At the end of this article is a very full list of references.

All the photographs relating to the cosmochronotrope have kindly been supplied by the Smithsonian Institution, Washington.

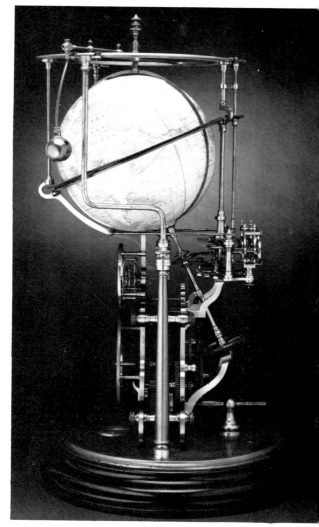

197c.

197d.

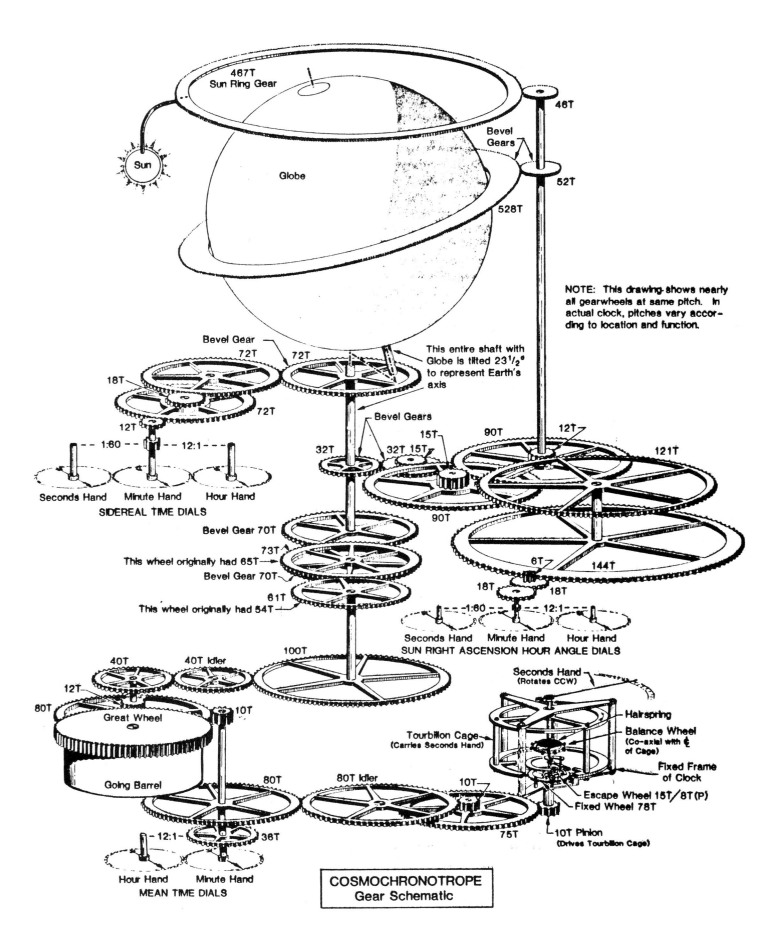

467T
Sun Ring Gear

Sun

Globe

Bevel
Gears

46T

52T

528T

NOTE: This drawing shows nearly
all gearwheels at same pitch. In
actual clock, pitches vary according to location and function.

Bevel Gear
72T

72T

18T

72T

12T

This entire shaft with
Globe is tilted 23½°
to represent Earth's
axis

1:60

12:1

Seconds Hand   Minute Hand   Hour Hand
SIDEREAL TIME DIALS

Bevel Gears

15T

90T

12T

32T

32T 15T

121T

90T

Bevel Gear 70T

73T

This wheel originally had 65T

6T

144T

Bevel Gear 70T

18T

18T

61T

This wheel originally had 54T

1:60

12:1

100T

Seconds Hand   Minute Hand   Hour Hand
SUN RIGHT ASCENSION HOUR ANGLE DIALS

40T

40T Idler

12T

Seconds Hand
(Rotates CCW)

80T

Hairspring

Great Wheel

10T

Tourbillon Cage
(Carries Seconds Hand)

Balance Wheel
(Co-axial with ¢
of Cage)

Going Barrel

80T

80T Idler

10T

Fixed Frame
of Clock

75T

Escape Wheel 15T/8T (P)
Fixed Wheel 78T

12:1

36T

10T Pinion
(Drives Tourbillon Cage)

Hour Hand   Minute Hand
MEAN TIME DIALS

COSMOCHRONOTROPE
Gear Schematic

REFERENCES

Britten (15th Edition revised by J.W. Player) *Watch and Clock Makers' Handbook*. E. & F.N. Spon Ltd., London. 1955.

Bruton, E. *Clocks & Watches*. Hamlyn, Feltham, Middlesex. 1968.

De Carle, D. *Watch and Clock Encyclopaedia, 2nd Edition*. N.A.G. Press Ltd., London. 1959.

Dreppard, Carl W. *American Clocks & Clockmakers*. Charles T. Bronford Co. M.A., U.S.A. 1958.

Drost. *Clocks & Watches of New Jersey*. P. 86.

Gibbs, J.W. "American Skeleton Clocks". *Bulletin. The National Association of Watch and Clock Collectors Inc.* Vol. XX No. 3. June 1978.

Kennedy, M. "The Banjo of Simon Willard." *Clocks*. Model & Allied Publications, Hemel Hempstead, Herts. Aug. 1983.

Lloyd, H.A. *The Collector's Dictionary of Clocks*. Hamlyn for Country Life Books. 1964.

Palmer, B. *The Book of American Clocks*. MacMillan, New York. 1928.

Palmer, B.A. *A Treasury of American Clocks*. MacMillan, New York. 1967.

Passmore, M. "Mass Production and the American Clock." *Clocks*. Model & Allied Publications, Hemel Hempstead, Herts. Aug. 1983.

Shelley, F. "Aaron Dodd Crane, An American Original," N.A.W.C.C. Inc. Columbia, Pennsylvania. Bulletin number 16. Summer 1987.

Terwilliger, C. "The 400 Day Clock and its American Connections." *Clocks*. Model & Allied Publications, Hemel Hempstead, Herts. Aug. 1983.

Tyler, E.J. "American Clocks in the pre-factory period". *Clocks*. Model & Allied Publications, Hemel Hempstead, Herts. 1983.

Willard, J.W. *Simon Willard and His Clocks*. 1962.

Bailey, C.H. American Clock & Watch Museum. Personal Communications.

7

Spain

It has only proved possible to trace one Spanish Clockmaker who made skeleton or skeletonised clocks and that is Manuel Gutierrez. The exact date of his birth is not known but it is believed that he was born in Siguenza a little prior to 1740.

In 1770 he applied for the Directorship of the School of Watchmaking which King Carlos III rd was at that time trying to form in Madrid. However he was unsuccessful in this attempt as indeed he was with many of the other important posts he applied for during his lifetime, possibly because of his temperament.

In 1789 he was awarded the contract for the provision of a clock for Toledo Cathedral (Fig. 198) and this he completed in 1792. Unfortunately there was some dispute as to the value of the work carried out which led to litigation, three experts from either side finally agreeing on a figure of 400,000 reales to cover Gutierrez fees, the salaries of those who assisted him in the installation of the clock and the various other expenses involved. Indeed by the time the council had paid all the additional expenses such as the stucco work and ornamentation of the ceiling; flooring, stairs, balustrades etc. the total cost was 727,751 reales.

A true skeleton clock which is unsigned but has been attributed to Gutierrez by Senor Luis Montanes[1] is shown in Fig. 12. It will be seen it has marked similarities to those being produced in Austria at that time.

Fig. 198. *The Toledo Cathedral Clock.*

This clock was commissioned as part of the great improvements being carried out in the Cathedral by Cardinal Lorenza at that time and was finally installed in 1792. It has interior and exterior dials and strikes the hours and quarters. The frame consists of twelve finely worked columns and bears allegories of the seasons of the year. Surmounting it and not seen in this picture is Cronos, the God of Time. The quality of finish is very high and all the bronze components, including the beautifully executed hands are firegilt. It is signed Dn. Manuel Gutierrez, Natural De Siguenza. Reloxero De El Rey F. EN MADRID.

Fig. 199. Three train skeleton clock from the Martinez
Jewellers which has a fully skeletonised movement and
dial. Although unsigned it has been attributed to
Gutierrez.

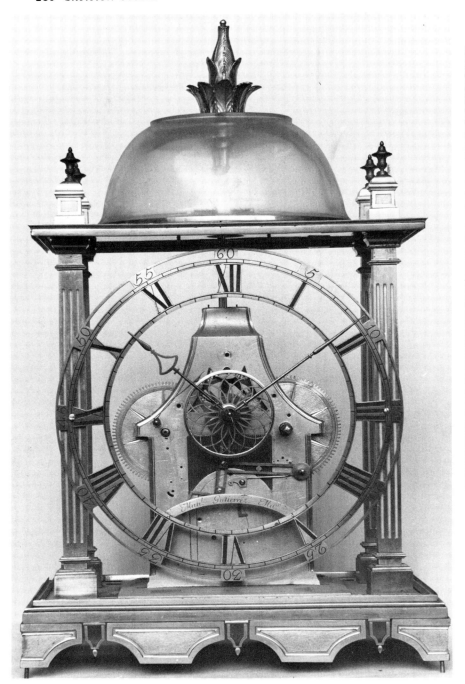

Fig. 200.

Fig. 201.

Figs. 200 & 201. Two very similar clocks, one of which was made by Gutierrez for King Charles IV and is now in the Madrid Palace (Fig. 200), and the other made for the Infante Don Luis.

Both glazed cases are made of polished steel with ormolu mounts. The 8-day movements have verge escapements with "silent" pallets and both originally had grand sonnerie striking, but sadly this has been removed from the King Charles clock. The power for the strike work is provided by an auxilliary spring that is automatically rewound from the single large spring barrel

which is provided with a fuzee.

The clock has a jumping hour hand which moves at the same time as the strike, the bell for which is concealed in the base. As with all clocks which have a remontoire spring for the strike a problem arises if this spring runs down whilst resetting the hands of the clock. To overcome this the strike/chime train may be stopped so that when the hands are moved the clock does not strike and thus run down the spring.

The clock in the Royal Palace (Fig. 200) has its original dome and finial, but those on Fig. 201 are replacements.

REFERENCES

Montanes, Luis. *Manual Gutierrez, Un profesional a ultranza.*
Montanes, Luis. *Relojes Espanoles*, ed. by Prensa, Espanola, 1968.
Montanes, Luis. *Anales Toledanos*, 11, Toledo, 1968.
Montanes, Luis. *Revista Iberjoya*, 3, Madrid, 1981.

Fig. 202. This shows the skeletonized movement of a quarter chiming 8-day longcase clock by Gutierrez made almost entirely of steel. It measures 50 x 40 x 20 cms. The dial is rectangular and pierced out as in the other clock. Photo Patrimonio Nacional, Madrid.

In addition to the clock mentioned Gutierrez made two double sided fully skeletonised gold watches numbered 1 and 2 and a metronome dated Madrid 1820 which was fully skeletonised and was designed for the charging of the fuses of grenades.

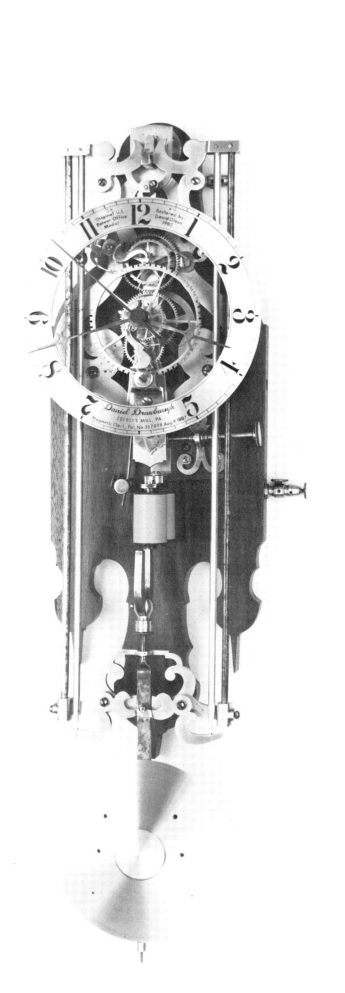

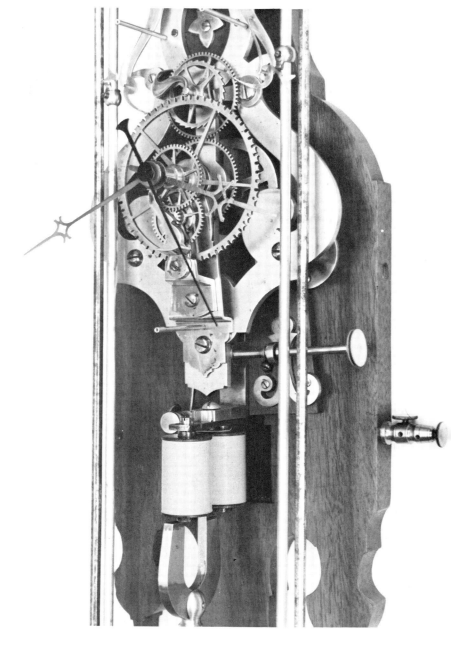

Figs. 203a, b. This clock, which would originally have been driven by an earth battery, made by embedding plates of copper and zinc in the ground, is signed "Original U.S. Patent Office Model" & "Daniel Drawbaugh, Eberly Mill P.A., Magnetic Clock, Patent No. 367898 Aug. 9th 1887". It was restored by David Olson of Burbank California in 1982 and it is he who kindly supplied the photos. Only three such clocks were built.

The coils pull the pendulum to the right when the sliding bar at the top of the pendulum makes contact in that direction; then the current shuts off as the pendulum returns to the left, the sliding bar being pushed by an insulated piece, so no contact is made. The large wheel with Geneva stop-watch type teeth is advanced one tooth each minute by a pin on the rachet wheel arbor. At all other times it is locked.

8

Electric Clocks

Right from their earliest days the majority of electric clocks have been made so that all the details of their movements could be seen as completely as possible. They were frequently displayed under a glass dome, glass dials were often employed or the dial centre was omitted and the frame, if such it can be called, was kept as open as possible.

Whilst it is impractical to go into the subject of electrical horology in any depth in a book such as this it was thought that readers might like to see pictures of a small selection of electric clocks so as to appreciate just how close they are to skeleton or skeletonised clocks in their overall conception.

Figs. 204a, b. Mid-19th century mahogany cased electric regulator with 5 rod gridiron pendulum and glass dial signed by F. de Jong of Amsterdam, whose apprentice piece is shown in Fig. 182 of the section on Dutch skeleton clocks. Photos taken by courtesy of the Curator, Clock Museum, Schoonhoven.

Figs. 205a, b, c. An interesting electric skeleton clock similar in overall design to the spring clock seen in Fig. 47. It is signed on the enamelled dial Paul Garnier, Hger De la Marine and is also signed on the backplate. It has a mottled marble base; wood rod pendulum; centre sweep seconds, minute and hour hands and runs on 1.5 volts.

The pendulum receives an impulse and the second hand advances once a second. The minute hand is synchronised but the hour only advances every quarter. The clock would date circa 1860. Height: 27½'' (57 cms.). Photos and details courtesy Mario Crijns, Breda, Holland.

Fig. 206. This skeleton timepiece which has a wood rod
pendulum with an elliptical bob and adjustable pallets to
the dead beat escapement is interesting in that the
mainspring is automatically rewound at short intervals by
a low voltage electric motor; which thus acts as a
remontoire.

This type of clock is discussed in some detail in Alan &
Rita Shenton's *The Price Guide to Clocks 1840-1940, 2nd
Edition*, pages 390-91. Suffice it to say here that an
electrical rewind mechanism was patented by Chester
Pond of New York in 1884 and further patents were taken
out by Max Hoeft in 1899 and by Max Moeller, Altona,
Germany between 1901-1907. Photo courtesy Sotheby's,
Bond Street.

Fig. 207. A skeletonized electric timepiece inscribed "Electric Clock made by The Reason Mfg. Co. Ltd., Brighton, Murdays Patent". It has a large horizontal balance and a form of single roller lever escapement. Height: 12½'' (32 cms.) Photo courtesy Sotheby's, Bond Street.

Further Reading

Aked, C.K. *Electrical Timekeeping*. Antiquarian Horological Society. 1976.

Aked, C.K. *Electrifying Time*. Antiquarian Horological Society. 1976.

Aked, C.K. *All you ought to know about the bulle clock*. 1983. Reprint.

Bain, A. *Alexander Bains Short History of the Electric Clock*. 1852. Edited by Hackman and reprinted 1973.

Hope-Jones, F. *Electrical Timekeeping*. 1970 reprint. Electric Clocks 1950.

Langman, H.R. & Ball, A. *Electrical Horology*.

Shenton, Dr. F.G.A. *The Eureka Clock*. 1979.

Wise, S. *Electric Clocks*.

9

Modern Skeleton Clocks

Although many collectors will object to the inclusion of modern skeleton clocks in this book, this policy has been quite deliberately adopted with the dual intentions of a) showing that modern craftsmanship is by no means dead, b) encouraging craftsmen both amateur and professional to produce further skeleton clocks so that we may at least have something to pass on to future generations.

The production of skeleton clocks on a commercial basis petered out soon after the turn of the 20th century but their construction by amateurs never completely ceased and since the end of the second world war interest in this type of clock has rapidly gathered momentum, particularly in the U.S.A.

The clocks being produced today may be divided into several groups.

1. Those made from kits.

2. Copies of 18th and 19th century designs. These may be made on an individual basis, usually by amateurs, or produced commercially.

3. Individual Designs. These fall into three groups:—

    a) Those produced commercially, often as a limited edition.
    b) Those commissioned from a professional clockmaker.
    c) Those produced by amateurs.

4. Skeleton Clocks made up either in total or in part from existing clock movements.

Clocks made from Kits

The most basic of these is undoubtedly constructed from Meccano (Fig. 208), the inclusion of which in a book of this nature, containing as it does so many masterpieces of the clockmakers art, will undoubtedly send a shudder down the spine of serious collectors and craftsmen. However their designs often show considerable ingenuity; they illustrate very graphically the vast variation in the range of skeleton clocks being produced and they have kindled an interest in horology in many people which has often led on to the production of far finer pieces.

Designs and instructions on how to make a skeleton clock have appeared in several journals over the years such as John Wildings articles on "How to make a skeleton clock" (Fig. 209) and "How to construct a Scissors Clock"[1]. In most instances these designs are backed up by commercial firms who will supply the majority of the components either in the rough or completely finished and thus it is up to each individual as to how much or how little of the work he wants to do. Many people, for instance, will not be able to cut their own wheels or pinions but can finish them and can cut out the plates, dials and hands. Few have a talent for engraving and thus tend to get this done for them, either using hand engraving or, if cost is a factor, have them acid etched. However dedicated the enthusiast there is nearly always something he cannot make such as a mainspring, jewels, or a hairspring.

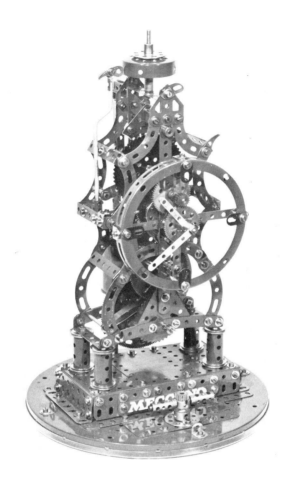

Fig. 208. Skeleton clock constructed entirely of Meccano by Mr. P.D. Briggs of Nottingham who has some 17 other examples made in the same way ranging from a weight driven lantern clock which has been going happily for some 15 years to an extremely accurate astronomical clock. There are now many Meccano Clubs in the U.K. and clocks are one of the items most commonly exhibited. A skeleton Hipp clock was described by Dr. Keith Cameron in the Meccano Magazine. Vol. 65 No. 1 Oct. 1980. Photo courtesy B.N. Love.

---

[1] Wilding, J. *How to make a Skeleton Clock* and *How to construct a Scissors Clock*. The Watch & Clock Book Society Ltd., Brant Wright Associates, P.O. Box 22, Ashford, Kent, England.

Fig. 209. A skeleton clock based on John Wildings design which was made by W. Smith, Knoxville, Tennessee. It won a Gold Medal in the Most Unusual category at the N.A.W.C.C. National in June, 1982.

There are also firms who design and market their own skeleton clocks[1], which may be bought either as a complete kit with finished or unfinished components, through to a fully assembled and even a finished clock.

Copies of 18th and 19th Century Designs.

Individually made Clocks. Like the clocks already mentioned some of the 18th and 19th century designs (Figs. 210, 211) have also been described in detail in various magazines so that the amateur can make them himself[2] and quite frequently the original designs are modified a little so as to make them easier for the amateur to construct. However on occasions an enthusiast may decide to make such a clock without any such assistance and in some cases the clocks may be commissioned from a professional clockmaker by someone who wants a specific design, such as that made by Tim Brameld in 1981 (Fig. 212).

The two fine clocks which were made by George Daniels in 1968 when he was London Agent for Breguet (Fig. 213) are in a somewhat different vein. In accordance with the tradition of the firm they were examined by the last of the Browns to run Breguet, who being entirely satisfied with their quality, issued them with Breguet Numbers.

One of the most ambitious copies of an 18th century clock ever undertaken was Mr. L. A. Salzers full reproduction of Harrisons No. I (Figs. 214a, b). The project was first conceived in 1965, commenced in 1969 and finished in 1977 some eight years later. The determination and dedication required of a job such as this are monumental, particularly as in this instance Mr. Salzer made virtually everything himself, including the balance springs which required a tremendous amount of research and trial and error. He even cast the spandrels up on his barbeque and made the chain, his comments on which are somewhat sobering "The chain was a *little* tedious to make but done over short periods it was not too hard on the eyes. There is 6 ft. of fuzee chain in the clock and I could assemble one inch an hour, not including the time spent on making the links".

It's amazing how in his brief article[3] Mr. Salzer dismisses his efforts so lightly even though the time and thought given to it was obviously so great.

Commercial Reproductions. Over the years several firms such as Thwaites & Reed and Dents have produced limited editions of various 18th and 19th century clock designs, frequently skeleton and novelty clocks such as Dents epicyclic; rolling ball (Fig. 215) and Whitlock Eagle Clocks (Fig. 216).

Individual Designs

a) *Those produced commercially, frequently as a limited or small edition.* These clocks (Figs. 217-222) often possess considerable artistic merit and are usually well made by men who would consider themselves as the natural successors of the clockmakers of previous centuries.

b) *Those commissioned from clockmakers on a "one off basis",* although this does on occasions result in 3-4 similar clocks being made. In this group are some particularly fine skeleton clocks which tends to confirm our feeling that excellent craftsmanship is still available, albeit in limited amounts and it is the demand, as much as anything else, which is holding back the production of more fine pieces. This is why we should be so grateful to collectors such as the late Major Heathcote who commissioned several fine skeleton clocks. (Figs. 223, 224). Undoubtedly makers such as Sarton, for instance, would not have produced the superb clocks they did without the patronage and help of wealthy clients.

[1] Classic Clock Kits, Dept TC/1 75 Foxley Lane, Purley, Surrey. CR2 3HP and J.U.B. Deckert, Suitbertusstr. 151, 4000 Dusseldorf-Unterbilk, W. Germany.
[2] Wright, J.G. "Joseph Merlins Band Clock", *Timecraft* 1985, Brant Wright Associates, P.O. Box 22, Ashford, Kent, England.
[3] Salzer, L.A. "The Making of a full size model of No. 1". *Horological Journal*, March 1983.

c) *Clocks made by Amateurs*

The term amateur is all too often used to imply that something has not been made or restored to the highest standards. However in the field of horology this is frequently very far from the truth. Because the amateur does not usually have to produce something within a set period of time or at a particular cost he can often take far longer making and finishing it than his professional colleagues. Moreover he can afford to make something because it appeals to him whereas the professional can usually only make a clock which appeals to his customers and when it is a highly specialised piece, has been commissioned from him.

It is in the field of clock design that the amateur often comes to the fore, good examples being (1) The 4-minute Tourbillion Clock made by David Olson in 1977 (Fig. 231) which took approximately 1000 hours to make and won The Gold Medal in the Masters Competition of the N.A.W.C.C. In 1980 (2) The skeleton clock by W. R. Smith with his own form of two wheeled grasshopper escapement and compound pendulum (Fig. 229) and (3) Dr. Curtis Fosters Clock (Fig. 230) with its' intriguing and complex pendulum.

It is to the credit of associations such as the Worshipful Company of Clockmakers, the N.A.W.C.C. and the B.H.I. that they organise competitions which stimulate and encourage such craftsmanship.

Skeleton Clocks made up in whole or in part from existing movements.

Quite frequently clock movements come to hand which are of little use as they are because they have lost their case and also sometimes their dial and even various other components. In these instances the first thought is often how they can be made use of and one of the easiest and best answers is in the construction of a skeleton clock. Frequently all that is required is to make a pair of frames between which the train or trains of wheels can be laid out in the most attractive fashion; fit a dial and find a base and dome. A good example of this is shown in Fig. 234 in which the wheels of a dial clock movement have been housed in a frame based on Arbroath Abbey. The only disadvantage of this approach is that skeleton clock movements, because they are on display, are normally made to a higher standard than those of bracket and wall clocks.

A far more elaborate approach (Fig. 235) is the construction of a 3-train skeleton clock from an old skeleton clock which was purchased in a scrap state. In this instance the design chosen was quite elaborate and thus many new components had to be made.

The year duration great wheel skeleton clock seen in Fig. 236 is largely of the clockmakers own making. However because of the high standard of most French movements he has made use of components from one, such as wheels and pinions so as to reduce the time taken in the construction of the clock.

At the end of this chapter we felt it appropriate to show a couple of skeletonized watch movements to remind ourselves just how fine the work of our colleagues in this field can be. (Figs. 238 and 239).

Fig. 210. A copy of Joseph Merlin's band Clock made by John G. Wright and described by him in *Timecraft*[1]. It gained a Silver Award at the 1983/4 Model Engineers Exhibition at Wembley.

The clock is the same size as Merlins original masterpiece but detail changes have been incorporated, some to improve the design and others to simplify its' construction. Examples of the former are the moving of the barrel ratchet so as to make it easier to set up or let down the mainspring and the provision of an adjustable crutch. Examples of the latter are the modification of the design of the finials so that they can be turned out of the solid rather than cast and the use of acid etching for the barrels and dial, rather than hand engraving.

The only components not made by Mr. Wright were the mainspring, the hairspring for the pivoted arm, the jewels and the glass dome.

[1] Wright, J.G. "Joseph Merlins Band Clock", *Timecraft* 1985, Brant Wright Associates, P.O. Box 22, Ashford, Kent, England.

Fig. 211. An epicyclic skeleton clock based on Strutts original design. It was made by W.R. Smith of Knoxville, Tennessee, U.S.A. and won the gold medal in the category for "Excellence of Workmanship" at the N.A.W.C.C. National convention in June 1982. The main differences from the original clock are that lantern instead of cut pinions are used and a cylindrical bob has been substituted for the lenticular one. A nice touch is the key with the makers initials incorporated in it.

This is a design which has been copied many times during the last century both commercially and by amateur horologists. E. Dent and Co. produced just such a clock in the early 1970s. Its attraction obviously lies in the technical difficulties to be overcome in cutting epicyclic wheels. Height: 9" (22 cms.)

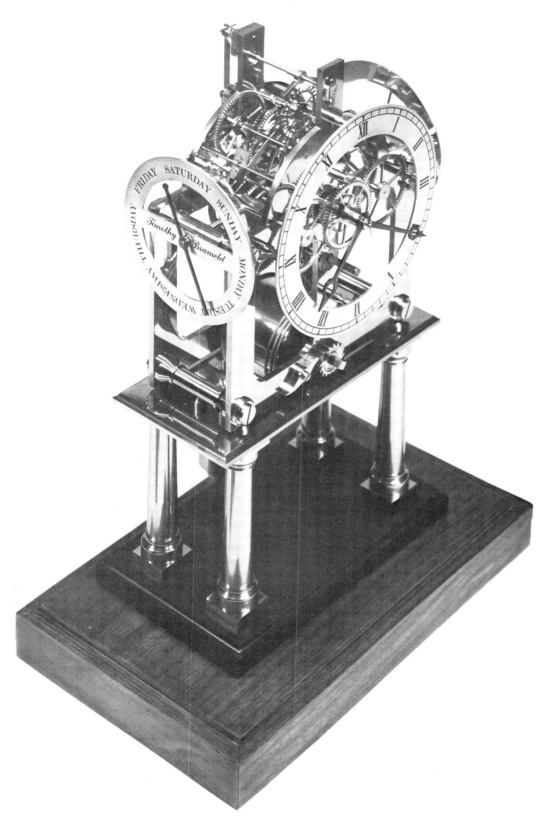

Fig. 212. This clock was based on one which belonged to the Northern Lighthouse Authority and is now in the Royal Scottish Museum, Edinburgh. The original spent its working life in the Cape Roth lighthouse.

It was made on commission by Mr. T. Brameld of Edinburgh in 1981 who has also made copies of James Edwards skeleton clocks which have cut glass centers to their wheels.

Figs. 213a, b. Two of these beautifully made three wheel clocks were produced by George Daniels in 1968 when he was London Agent for Breguet and are described in his book on that maker[1]. In accordance with the traditions of the firm M. George Brown, lately retired proprietor and the last of the Browns to run this business examined the clocks and being satisfied with their quality entered them in the Companys book allocating them the numbers 3224 and 3225 and issuing them with certificates; that shown here being No. 3225.

The bases of the clocks were copied from one of the original three wheel clocks and the engraving was carried out by Allen Lye principal engraver to the Bank of England using a photograph as a guide.

These clocks, referred to as "Three Wheeled Clocks", have a pin wheel escapement with a half seconds pendulum supported on a knife edge. Compensation is effected by means of a bi-metallic strip above with a scale indicating the change of temperature. The driving weights turn a single arbor which carries the wheel for the hours which is calibrated 1 - 12 twice in Arabic. The pointer on the right hand column indicates mean time and that on the left is for the equation.

There is a scale for the minute with three armed pointer mounted on the front plate immediately above the hour ring and a dial below it gives the Julien and Revolutionery calendars. Engraved on the front frame are the days of the week and these are indicated by the level of the tops of the weights.

---

[1] Daniels, G. *The Art of Breguet.* Sotheby Park Bernet. 1975.

Probably the most famous skeletonized clock ever made was Harrison's Sea Clock No. 1 which was eventually to lead to mans ability to record time accurately at sea and thus determine the exact longitude and make possible the production of accurate maps. The original clock may be seen still ticking away quite happily and most impressively in the National Maritime Museum in Greenwich. Its' movement is too complex to describe in detail here but for an overall account of its' construction one can do not better than quote from Commander Gould[1] who was responsible for its resurrection in the 1920s and 30s and restored to it the magnificent state we now see it in.

"It is really a large marine clock, controlled by two huge straight-bar balances mounted on portions of large anti-friction wheels, and connected by cross-wires running over brass arcs. They swing, in consequence, as if geared together (but with far less friction) and a ships' motion has no appreciable effect on their period of oscillation. They are controlled by four helical balance-springs, in tension; and a triple "gridiron" of brass and steel rods automatically varies the tension of these springs so as to counteract the effect of heat or cold on the springs themselves. This is the first compensation for temperature ever applied to any time-keeping instrument controlled by a balance.

The wheels (except the escape wheel, which is brass) are of wood (oak) with the teeth, also of oak, morticed into the rims. They are all mounted on anti-friction wheels, and move with remarkable freedom. There is no remontoire. Two main-springs drive a single central fusee, provided with a "maintaining spring" to keep the machine going while being wound (which was, originally, accomplished by pulling a cord wound on the fusee itself). There are two escapements, of the "grasshopper" pattern, one being mounted on each balance-staff. The machine goes for about 38 hours at one winding, and shows seconds, minutes, hours and days. For use at sea, it was enclosed in a wooden case, suspended by springs from a gimbal-frame."

[1] Gould, R.T. *John Harrison and His Timekeepers.* National Maritime Museum 1958. HMSO Dd 559231, K 24 11/73 3309.

Figs. 214a, b. Mr. L.A. Salzers' superb copy of Harrison's machine, which is now on loan to the National Maritime Museum.

Because he could not actually examine and measure the components of the clock he photographed it from all angles so as to assess their relative dimensions and then made a cardboard model, painted this white on the edges and photographed this from the same positions as the original. He then compared the two sets of negatives and adjusted the model sizes accordingly, thus obtaining the final frame sizes.

To describe all the work involved in constructing a clock such as this could virtually take another book, suffice it to say here that over a period of 8 years Mr. Salzer gradually overcame all the problems, including the escapement, the wooden wheelwork, the pinion rollers, finding a mainspring which matched the profile of Harrisons original fuzee, constructing a chain, making the correct temperature compensation and assessing and making the balance springs. The dedication required by projects such as this is illustrated to some degree by one of the clockmakers remarks "The chain was *a little* tedious to make but done over short periods it was not too hard on the eyes. There is 6 ft. of fuzee chain in the clock and I could assemble 1" in an hour, not including the time spent on making the links."

Having completed the first clock it says much for Mr. Salzers' tenacity that he promptly started on a second one for the Time Museum in Rockford, IL., U.S.A.

Fig. 215. Congreve's rolling ball clock has been copied on numerous occasions by clockmakers, both amateur and professional, since its original conception in the early 19th century. The example shown here was made by Dents in the 1970s but other firms such as Thwaites and Reid and Bell of Winchester have also made similar clocks. It has to be made particularly well if it is to perform reliably, especially with a table which tilts every 30 seconds as any slight inaccuracy in the track may well result in the clock stopping. Difficulty will also be experienced if the ball and table are not kept meticulously clean.

One way of reducing this problem is by employing a table which tilts every 15 seconds, thus having a ball which runs faster and is less effected by any minor inaccuracies.

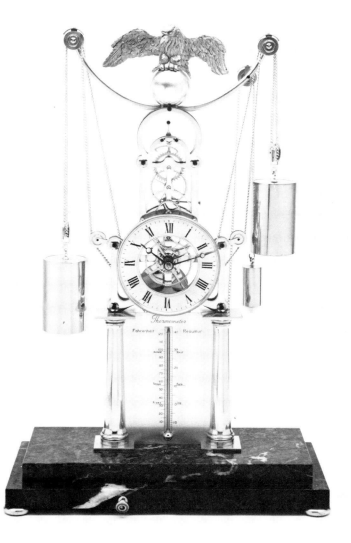

Fig. 216. A weight driven skeleton timepiece with anchor escapement marketed by E. Dent & Co. Ltd., in the 1960s under the title "The Whitlock Eagle Clock". It was based on a design originating in France in the 1840s. Only one of the larger weights actually drives the clock, the other being a dummy. The chain is wound onto the barrel by pulling a cord in the base.

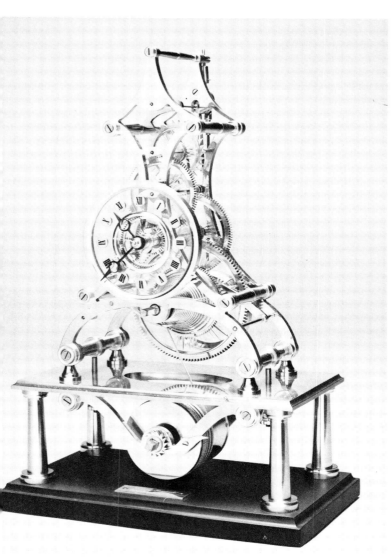

Fig. 217. A skeleton clock produced in limited numbers by Timothy Brameld of Edinburgh. It was inspired by those made by James Edwards of Stourbridge which have decorative glass centres to the wheels and pendulum bob and incorporates maintaining power.

Fig. 218. One of a limited edition of 100 great wheel skeleton clocks made by Varley Brothers of Norwich, England. This attractive clock which is of fine quality stands some 23'' high. The 8 day movement employs a 400 tooth great wheel, has enamelled numerals and hands; shows moon phases and has a centre sweep date hand. It stands some 14'' (35 cms.) high.

Fig. 219. A triple framed double great wheel clock with the mainspring wound from the same arbor and count wheel strike. Dead beat escapement and a half seconds pendulum are employed. This is one of a range of good quality clocks produced in limited numbers by David Walter of Bedford Park, Western Australia.

Fig. 220 A well constructed 8-bell skeleton clock by Sinclair, Harding and Bazeley of Cheltenham, England who produce an attractive range of hand crafted clocks mostly to their own designs.

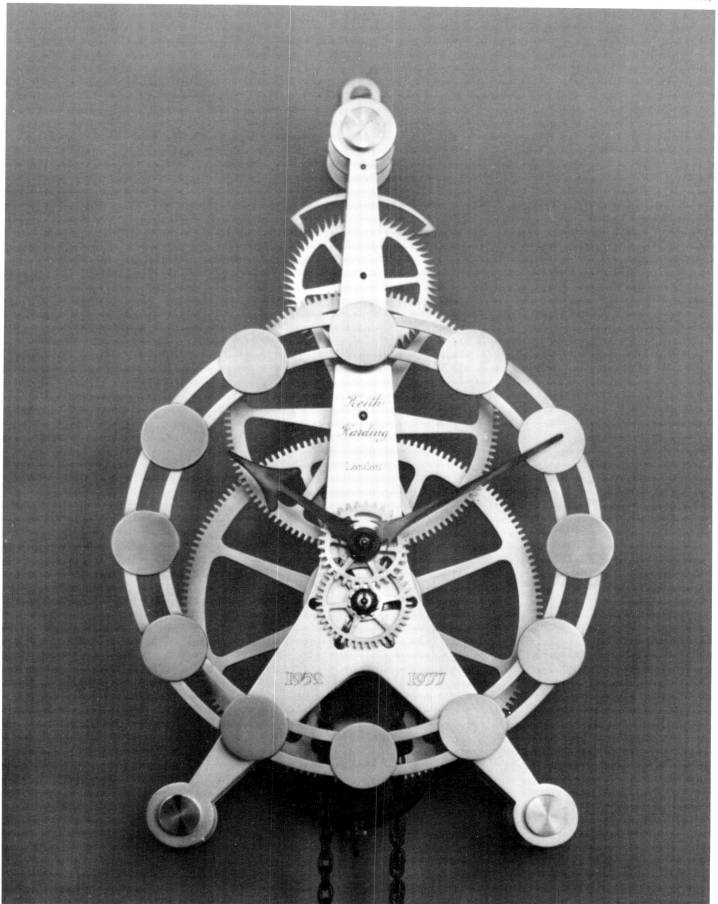

Fig. 221. A timepiece designed and made by Keith Harding in 1977 as a limited edition of eight, when it formed part of a special exhibition of work by contemporary British Craftsmen in the Victoria and Albert Museum to celebrate the Silver Jubilee of Queen Elizabeth II.

Figs. 222a, b, c. A skeleton clock of attractive design and considerable ingenuity made in limited numbers in the late 1960s by a gifted clockmaker who prefers to remain anonymous. Sometimes he left the clocks unsigned and in other instances used a pseudonym.

The dial layout is particularly interesting with separate overlapping rings for the seconds, minutes and hours. A pin wheel escapement is used with "Coup Perdu" and thus seconds are shown on the dial although a half seconds beating pendulum is employed. The compensation used on this is somewhat similar to that devised by Ritchie in that it makes use of a bimetallic bar. This extends from either side of the bob and has adjustable weights attached to the ends of it. As the temperature changes the bar either increases or decreases in curvature and thus raises or lowers the auxillary bobs. By moving these in or out the degree of compensation may be varied. The overall view and that of the escapement are taken from a clock owned by Dr. Marsh of North Haven, Connecticut, U.S.A. and the close-up of the pendulum is by courtesy of Mr. Heldman of Birmingham, Michigan, U.S.A.

Figs. 223a, b. A precision skeleton clock commissioned by Major Heathcote in the early 1970s. The silvered brass chapter ring, which has a skeletonised centre, incorporates a year calendar showing the days, months and the corresponding signs of the zodiac. The free sprung spring detent escapement, which is fascinating to watch, employs a large helical hairspring mounted above the massive balance wheel some 5'' (12.5 cms.) in diameter. Maintaining power is provided. Height: 15'' (38 cms.). Photographed by Kind Permission of Sothebys, Bond Street.

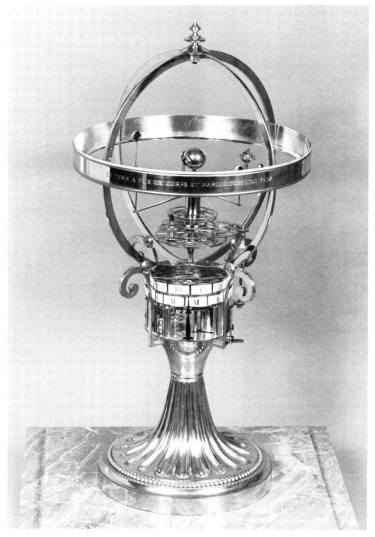

Figs. 224a, b. An Orrery Clock made for the Heathcote collection. Rotating rings with enamel chapters indicate the minutes and hours. The movement has a horizontally mounted lever escapement and a going barrel to which the astronomical work is geared. This shows the relative motions of the Earth, Moon and five planets, each being depicted by coloured jewels. Surrounding this is a ring inscribed on the outside "Le Tems a pris un Corps et Marche sous nos Yeux" on the inside is a year calendar incorporating the signs of the Zodiac. Photographed on their premises by kind permission of Sothebys, Bond Street.

Fig. 225. An ingenious 8 day skeleton clock in the form of Cleopatra's Needle made by Peter Bonnert of Maidstone in 1984 and numbered 109. The train wheels and pinions are of very high count and the chapters are engraved on the rims of the wheels and thus rotate, time being indicated by fixed pointers. The motive force is delivered by a steel 'crossbow' drawn into tension by two small chains acting upon a drum fixed to the great wheel arbor. The crossbow is recocked every 2-1/2 hours by means of a remontoire housed in the base. The centre wheel is provided with Harrison's maintaining power to keep the clock going during the recocking process. Because of the tapering shape of this clock the wheels are run in plates which are not parallel, which necessitates pivot holes which are not at right angles to the plates, and all the arbors are of different lengths. The escapement is a light form of pivoted detent chronometer and the 'scape wheel and detent pivot holes are jewelled. The clock is wound through a hole in the marble base which facilitates winding without removing the brass edged glass case. The frames and plinth are frost gilded.

Train count

Great Wheel 360		Centre Wheel Pinion . . .30
Centre Wheel 220		Third Wheel Pinion 24
Third Wheel 144		Seconds Wheel Pinion . .22
Second Wheel. 84		'Scape Wheel Pinion . . .12
Scape Wheel 8		Overall Height: 23".

Fig. 226. A fine modern table regulator produced by
Bernhard Lederer of Rodermark, W. Germany. The
classic simplicity of the clock is most appealing and the
engine turned silvered chapter ring, reminiscent of the
work of Breguet, has jumping seconds, minute and hour
hands. The 9 rod temperature compensated gridiron
pendulum has an adjustable screw for one second a day.

Fig. 227. A miniature 8 day two train skeleton clock made
by Richard Cox of Irving, Texas. It won first place in the
novelty class of the 1982 National Convention of the
N.A.W.C.C. Its overall height is only 7". It has a
horizontal compound pendulum with the makers own
form of knife edge suspension, 2 pointed steel studs
resting in concave jewels.

Figs. 228a, b. A highly individual skeleton clock made by Mr. Richard Cox of Irving, Texas, an engineering design draftsman, in the machine shop of Gerald Fendley. It won first place in both the Master and Most Unusual Sections in the 1979 N.A.W.C.C. convention in the U.S.A. Its height is 17''.

Interesting features include:-

Double dumb-bell type cross beating seconds pendulum.
A split grasshopper escapement to Mr. Cox's own design. This may be seen in the close up.
The fitting of both maintaining power and stopwork.
A high count train with very large diameter finely executed decorative wheelwork.
The concealing of the mainspring in the base.
A miniature copy of the main dial mounted immediately above it which shows seconds.
A well designed and finished scroll frame.

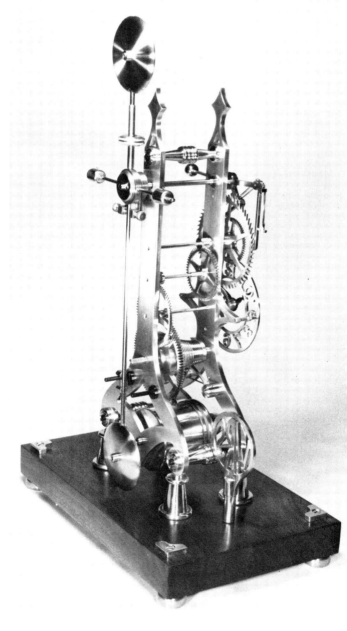

Figs. 229a, b. An 8 day skeleton clock made by W.R. Smith of Knoxville, U.S.A. to his own highly individual design. It incorporates an ingenious form of two wheeled grasshopper escapement devised by him. This has been cantilevered to the front of the front plate by the use of one ball bearing for the front of the 'scape arbor and a ball bearing at each end of the pallet arbor to allow the pendulum and pallet assembly to be mounted on the same arbor, a most interesting arrangement.

The compound pendulum some 17'' long beats 72 times per minute whereas a standard pendulum of this length would beat 92 times. The train count is: 'Scape wheel 60, Escape Wheel Pinion 10, Third Wheel 60, Third Wheel Pinion 12, Centre Wheel 72, Centre Pinion 8, Great Wheel 96.

The only commercial parts used in this clock are the mainspring, fuzee cable, ball bearings and screws. Overall Height: 19''.

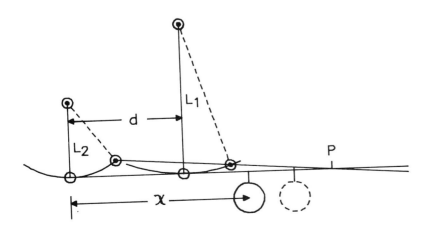

Figs. 230a, b, c. This highly individual clock, which was made by Dr. Curtis Foster of Barrington Hills, U.S.A., won the Gold Medal in its class (most complicated) and also the Master Gold Medal in the N.A.W.C.C. U.S. National Achievement Award contest in 1978. It incorporates a pendulum based on the interesting principle that it is the arc which the pendulum bob follows which determines the timekeeping, (except in the case of compound pendulums) and not necessarily the length of the pendulum although in practice these usually relate to each other.

In the pendulum devised by Dr. Forster, two rods of unequal length are employed with a hinged interconnecting bar at their bottom which extends to one side of the longer rod and supports the bob. In Fig. 26c L1 and L2 are the lengths of the two rods, in this instance 10'' and 5'' respectively; d is the distance between them and X is the distance between the bottom of the shorter rod and the centre of the bob.

As the pendulum assembly swings to one side the shorter rod will obviously go through a larger angle than the longer one and thus its' height above the stopped position will be greater. The connecting rod will now slope down to the right and if this is extended far enough there is a point at which the height of the bob will stay constant as the pendulum swings. This theoretically corresponds with a pendulum of infinite length. Thus in practice by varying the position of the bob along the rod its theoretical arc of swing can be changed from 10" to infinity.

The length of the equivalent conventional pendulum Le can be calculated by using the following formula.

$$Le = \frac{L_1 L_2 d}{d L_1 - x L_1 + x L_2}$$

for $L_1 = 10"$; $L_2 = 5"$; $d = 1"$ we get

$$Le = \frac{50}{10 - 10x + 5x} = \frac{10}{2 - x}$$

We can now see that when $x = 2"$ the denominator goes to zero and we have an infinite length. When $x = d = 1$, we are at the bottom of the 10" rod and have a 10" pendulum. To get 39.14" we have

$$39.14 = \frac{10}{2 - x} \text{ or } x = 1.745"$$

which is a realistic value to achieve.

The rate of the clock is controlled by a micrometer which moves the bob which weighs 2.5 lbs, horizontally on the rod. A point to bear in mind is that the suspension of the shorter rod will be in compression and this must be allowed for.

Other interesting features of the design are the frames, which are cut out in the form of the makers initials, and the gravity escapement. In this the pallets are truly dead beat and transmit no power to drive the pendulum; this function being carried out by two balls which are alternately raised 0.015" (.375 mm) and then released onto the anchor. (Fig. 230b). Photos courtesy Mr. David Stephans.

Figs. 231a, b. A four minute Tourbillion Clock which took David Olson of Burbank California some 1,000 hours to make. It was awarded the gold medal in the Masters Competition of the N.A.W.C.C. at Boston in 1980. The weight of the balls in the baskets drives the escapement. When a ball reaches the bottom it rolls into a hole in the cage, thus releasing the train, which revolves one space. This allows the ball at the top to roll into the basket and at the same time the minute hand is advanced one minute.

The Tourbillion is driven by two mainsprings working in tandem and is connected to a conventional French mid 19th century movement with count wheel strike on a bell.

The idea of the Tourbillion i.e. a carriage in which the escapement revolves, was invented by Breguet so as to try and overcome positional errors in watches. It was usually employed with a chronometer or, as in this case, a lever escapement. Towards the end of the 19th century B. Bonniksen invented a somewhat similar revolving carriage called a Karrusel which rotated every 51-1/2 minutes. Because of its' relatively slow speed of rotation it is not nearly as interesting to watch as a Tourbillion but is probably just as effective.

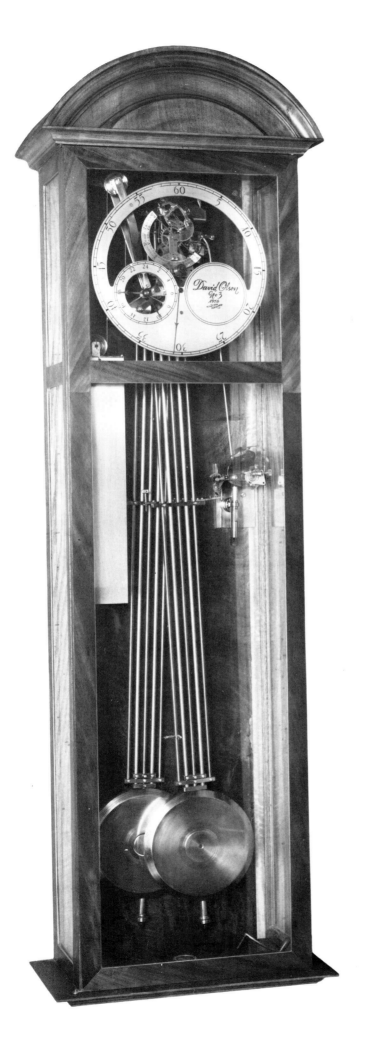

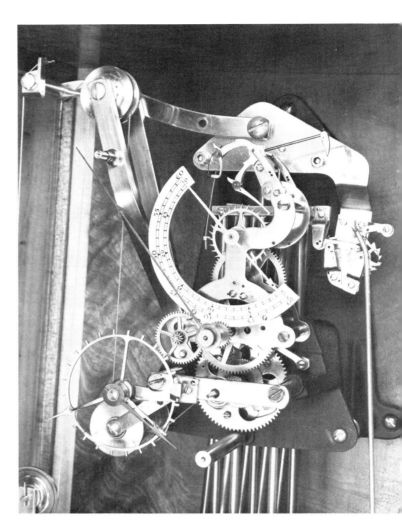

Figs. 232a, b. A fascinating skeletonized weight driven wall regulator made by David Olson of Burbank, California. It took the Gold Medal in its' category at the N.A.W.C.C. National Convention Competition in 1976.

It employs master and slave gridiron pendulums which swing scissors fashion. The grasshopper escapement drives the front (slave) pendulum which advances a count wheel each time it swings across and every 30 seconds this releases a roller which impulses the rear pendulum. The front pendulum raises the gravity roller by means of an interconnecting rod and at the time of release the rear pendulum compares the rate of the slave pendulum, which is adjusted to run slow, and accelerates it as necessary.

The dial has a centre sweep minute hand, hours are shown at the bottom left (local and Greenwich Mean Time) and seconds are indicated on a semi-circular strip graduated 0 - 30 and 30 - 60, with a double ended seconds hand.

The basic principle of the employment of Master and Slave pendulums is that one pendulum is allowed to swing completely free from any interference which the escapement may have on the regularity of its' vibrations.

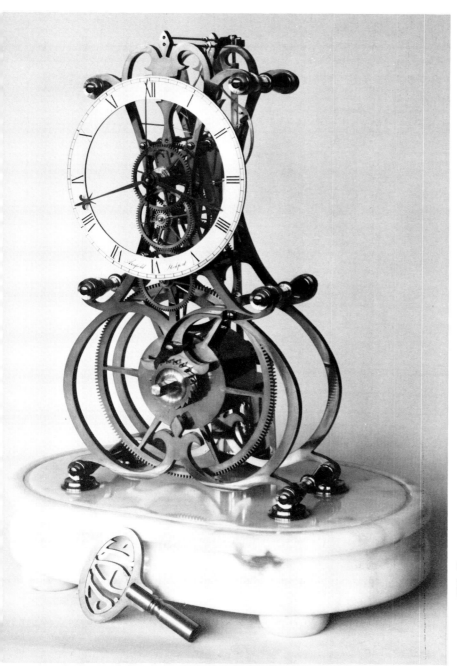

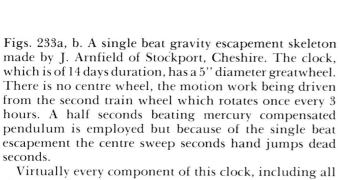

Figs. 233a, b. A single beat gravity escapement skeleton made by J. Arnfield of Stockport, Cheshire. The clock, which is of 14 days duration, has a 5'' diameter greatwheel. There is no centre wheel, the motion work being driven from the second train wheel which rotates once every 3 hours. A half seconds beating mercury compensated pendulum is employed but because of the single beat escapement the centre sweep seconds hand jumps dead seconds.

Virtually every component of this clock, including all the pins, screws and washers; the saphire locking stone and the marble base were made by Mr. Arnfield. The only components bought out were the mainspring and the glass dome.

Fig. 234. A skeleton clock made by Mr. G. Smith of Maulesbank based on Arbroath Abbey which was founded by King William (The Lion) of Scotland in memory of Thomas Becket, Archbishop of Canterbury. The actual part of the building depicted is St. Catherine's Window, popularly known as the Round O. The scalloped part of the frame represents the shape of the building. The wheelwork was obtained from an old dial clock. Photo courtesy Mr. Iain Wright.

Fig. 235. A three train skeleton clock quarter chiming on 8 bells and striking the hours on a gong. It was commissioned by Major Heathcote and made by Malcolm Blandford CMBHI in the workshops of Aubrey Brocklehurst in London in 1970. Some of the components including the dial were derived from an old skeleton clock which was purchased in a scrap state. Calendar work is incorporated above the main dial. Height: 1' 7'' (48 cms.)

Fig. 236. A Year duration Great Wheel fuzee skeleton
clock with the mainspring concealed in the base, made by
the same clockmaker as that shown in Fig. 222. It says
much for the quality of the original French movement
from which many of the wheels and pinions (excluding
the great wheel) in this clock were taken that it goes quite
happily for one year on a relatively small mainspring.

Figs. 237a, b. The basis of this clock is a standard 8 day spring driven movement. This was completely dismantled by Mr. Fritz Leitheim of München (who kindly supplied the photos), fully skeletonized and many of the components engraved prior to being reassembled. The movement was then modified and a rotating globe moon added.

Figs. 238 & 239. Although not really within the orbit of this book it was thought that the inclusion of two skeletonized watches would be of interest to many readers. Those shown were produced by Kurt Schaffo at Le Locle, Switzerland. Some ideas of the extreme delicacy of the work involved may be gained from the fact that one of the movements is shown fitted inside a coin.

Appendix

For an illustration of this clock which was supplied by Antiquorum, Geneva, see Fig. 107, page 115. The following text is a description of this book as extracted from *La pendule Neuchateloise* by Alfred Chapuis which gives a translation of an article which appeared in a German scientific review in 1819, and the clockmaker's story by A. Mongrel in which the secret of the clock's perpetual motion is revealed.

CLOCK WITH PERPETUAL MOVEMENT BY GEISER, UNIQUE PIECE, made in 1815 by JEAN AND DAVID GEISER, FATHER AND SON IN NEUCHATEL.
Who could better describe this marvellous clock than Alfred Chapuis in his researches about the Neuchâteloise clocks and M. A. Mongreul who wrote this amusing "true story" which I offer you here:

LA PENDULE NEUCHATELOISE

by Alfred Chapuis
Documents Nouveaux, page 239 and 240

Curiously enough, at the same time as the Maillardet, two other Neuchâtel clock-makers, Jean and David Geiser, to whom reference has just been made, displayed a so-called 'Perpetual Movement' of their own invention, from 1815 on. At first there were just vague rumours, but since then we have actually been able to look through a brochure in German which gave a very detailed description of this extraordinary machine. We have looked through the principal libraries in Switzerland and done everything within our power to find this brochure again, but in vain. However, an article which appeared in a major German scientific review in 1819 has been called to our attention, and we give the translation of it below[1]:

The Geiser Perpetual Movement

We have seen this machine, presented by Mr. Estermann, and we seriously recommend that our readers do not fail to take advantage of this opportunity to see it, for it merits being seen, from every point of view.

The form of the construction alone, the elegance of the workmanship, and the pendulum attached to the mechanism provide ample reason for paying homage to the creators of the work, Messrs Geiser, father and son, of Switzerland. Poppe[2] had the machine in his possession in Frankfurt for quite a while. He observed it minutely and his examination revealed nothing that might point to a

---

[1] Isis, oder Encyclopaedische Zeitung, 1819, p. 365-366. We thank the Count of Klinckowstroem, in Munich, who graciously supplied us with this precious information.

[2] Poppe was a scientific writer, very well-known in Germany during the first half of the 19th century.

hidden device. Indeed, we are not competent to judge this question, but we wish to state plainly that we are unable to imagine where any secret device could be concealed. In addition, when regarded, the machine appears to have no contact with anything else, and it is so ingeniously constructed in accordance with the laws of mechanics that no reason could be conceived why this movement should not continue perpetually.

The apparatus consists of a pierced brass wheel 1½ feet high. 24 cylinders (rollers) are attached to it, which can straighten out or fold back. When one of these cylinders is straightened, it lengthens one of the radii (levers): when it is folded back, this shortens the lever. The cylinders fold back at the top of the wheel, and at the bottom they straighten in such a way that, on one side of the wheel, the cylinders are always folded back while, on the other side they are always straightened out. Thus, additional weight is produced on one side. The cylinders are 1½ inches long and ¾th of an inch thick. The straightening-out and folding-back have been so ingeniously conceived that it would truly appear that this conception in itself would be sufficient to attain the goal sought for thousands of years—the mechanical realisation of the movement of life. Poppe has made a drawing of the apparatus.

There is only one possibility for dupery. A spring would have to be concealed in the arbor of the wheel itself, but this is hardly credible in view of the finish of the mechanism, which was able to keep the machine in operation for several weeks. And, furthermore, neither the appearance of the machine itself nor the personalities of its inventors justify the supposition of imposture. Naturally, nobody dares to cry from the housetops: "Eureka! Here is perpetual motion.' Everybody is fearful of believing that something has been brought into being here which has been proved impossible by the mathematicians. Despite this, we assert: "Go and see this machine for yourself. When Mr. Estermann comes to town, everyone should take the occasion to see him!

We are asking our readers to provide us with information on one point. It is permitted to touch the machine. In pushing the wheel back, one can very distinctly feel the resistance produced by the excess weight existing on the side where the cylinders are straightened. Without a doubt, those gifted with a delicately attuned sense of touch will be able to recognize whether this force is produced by a spring, for the latter could not fail to impart a certain impetus to the wheel. We were not able to reach a final opinion on this matter ourselves.

The Geiser clock, as we have seen, although it may not have provided the long-sought solution to the problem, was nevertheless, an exceedingly elegant mechanical construction. In the light of this, it is very strange that so little has been said of it in the region of Neuchâtel.

ONLY HUMAN STUPIDITY IS PERPETUAL

A true story
by A. MONGRUEL

This effort at perpetual motion, even if it did not end with its inventor in the madhouse, had the merit of intriguing more than one watchmaker, and particularly the last one who restored it, namely myself. And for good cause. I was young and still nourished many illusions. Everything conspired to arouse my enthusiasm and give me hope of being able to make the machine run *perpetually*, all the more so since my client, the owner of the clock, claimed to have seen it working in his own youth. He maintained that the connoisseurs admired it and that men of science came frequently to examine it. And, in addition, I did not believe that the watchmaking world knew about it, and that its description could ever have been made public.

When I received the clock it was in lamentable condition. The château in which it had been displayed in a place of honour in the salon had just been destroyed by a fire caused by lightning. The clock itself was not directly touched by the flames, but the firemen of the commune had sprayed it with diligence. And since it took some two months before the wreckage was cleaned up, the damage was all the more complete, and little was left of the clock other than a mass of rust and verdigris.

This clock, resting on a copper base, as the photograph clearly shows, contains two large crown wheels, both as fully perforated as possible to diminish their weight held apart in perfect parallel by pillars, in the same manner as the plates of a clock's movement. Eight arms, starting from a bronze hub of about 7 centimetres in diameter, pushed on out to the inner circumference of these crowns. The entire work was impeccably finished. The wheel which formed this ensemble was perfectly round, rotated without flaw, and was as exactly equilibrated as the balance of a watch. Twenty-four axes pivoted between the two circles which formed this wheel, and on these axes arms were adapted, at the end of which were fixed weights or masses each weighing between 400 and 500 grams, as well as a ratched wheel, two parts of a rack with pinions, one fixed at the bottom, on the frame or base outside of the wheel, and the other set in at the top, into the interior area of the circle. These had, as their function, when the ratchets reached them, to cause the axes of the weights to pivot, in order to make them return to the inside of the wheel, and inversely.

With this summary description, we can immediately recognize, as we look at the drawing, the ins and outs of the whole system. The movement of the pendulum and the rollers at the bottom (rear view) are only accessories, at least so far as the search for perpetual motion is concerned. Only the big wheel is of interest to us.

On the left side of the photo (rear view), we see the copper weights, exterior to the wheel—on the right side, these same weights are inside the wheel, brought into this position by the ratchet which is integral with them, the teeth of which are engaged with the rack at the base. These same ratchets later come into contact with the upper rack, which has the contrary function, namely to thrust the weights to the outside.

Taken individually, the exterior weights become an energy source in relation to the weights diametrically opposite which provide resistance and are a force to be overcome.

The levers to the left are much longer than those to the right and this provides the motor power which drives the apparatus. I no longer remember the exact measurements, but, by placing a ruler on the photograph, it can be seen that the difference in length of the levers is in the ratio of 5 on the motor side to 3.5 on the resistance side. The amount of driving force is thus very slight for evaluation.

Leaving aside the problems of resistance, of friction, of the weight of the wheels and the masses, there is no doubt that in this arrangement of the elements, a potency or a driving force is obtained and could be put to work.

but, unfortunately, there are those accursed factors of friction, of considerable weight to be displaced, and these constitute a very real embarrasment when you are looking for perpetual motion.

But that is not all: to the resistance of the large wheel itself must be added the very significant resistance of the clock movement itself, by the way, a most handsome one, with a circular pallet escapement, a dial train, and the hands to be driven. In addition, perhaps fearing an excess of energy capable of causing the escapement to knock, the inventor installed a speed governor to brake a possible racing of the wheels. This governor, seen at the bottom of the photograph (rear view), was composed of three very cleverly placed rollers, slowly and noiselessly rolling on a polished surface which, in this case, was a silvered glass.

Think it over, weight the evidence, draw you own conclusions. Keep in mind the heavy incrustations of rust, the pivots eaten away. There was no trace of a hidden driving force. Only the hub of the wheel, 6 or 7 centimetres in diameter, was protected from water by the frame and still looked like a mechanical part, lathe-turned with precision. In contrast, the pivot, extending out from each side, was red with rust. The hub, well-proportioned to the power and the diameter of the wheel, was about 25 millimeters thick, with a diameter of 6 or 7 centimeters.

There would be no point in listing the various substanced employed and the number of hours spent in restoring this pile of junk to beauty. Finally, at the end of the process, polisher's rouge and diamond dust replaced the oiled polishing stone, emery buffs and cloths, down to the finest, polishing pastes, etc. After three or four days of furbishing, everything was transported to a big work bench. This was the moment when tensions mounted and nerves snapped.

I was almost convinced that the wheel could turn, but what continued to bother me was the very limited amount of motor force that I could hope for. And I was troubled by the thought that, in addition to the resistance of the wheel, there was still the pallet escapement with circular balance wheel that had to be activated, and this, all by itself, needed more motor power than that required for maintaining the oscillations of a pendulum. But where I was completely lost was at the idea that I still had to keep in motion the governor that I have already described, especially since, when I tried it out separately by pushing the pinion at the end of its axis by hand, I encountered a rather strong resistance. But I was still full of hope, since my client had seen the clock in operation, and I began to assemble the whole thing, starting, of course, with the wheel.

Try to imagine the great care that was required, since this jumble of rusty iron, as I called it above, had become a gleaming display of the goldsmith's art, reflecting back the bright rays of the sun. The warm luster of the beautiful red bronzes at the base and feet of the clock, the polished steels—, all of that glowed with mirror brightness and would have filled me with delight, if it had not been for that furtive little apprehension that fluttered at the back of my brain.

We were almost tempted to wear gloves in order to make sure that we didn't leave a trace on the large, polished surfaces. But I had already requisitioned all of the cloth that could be found in the house that was soft enough and large enough to handle this great wheel, which measured 45 to 50 centimetres in diameter, and mounted on its base, stood at least three-quarters of a metre high.

Now that the wheel is set in its final position with all its weights and ratchets bizarrely toothed posed on its base, we make a last detailed inspection, checking the normal play of the axes between their bearings, making sure that it is impossible for the weights to impinge on anything in their revolutions, and then, the smallest possible drop of oil, applied parcimoniously to avoid its smearing, leaving unsightly trails across the beautiful polish—Now the fateful moment was at hand. Were we going to be recompensed for our patient efforts, for our hours of emotional anguish?

With the frame perfectly held in the horizontal position by the adjustment screws which can be seen in the photo, we set the wheel in motion, keeping a finger ready to brake it if it accelerated too quickly! And let us not forget that there was a governor to control the speed which gave rise to fears of blocking. At last we were rewarded. The wheel turned—one and a half times, approximately. The weights faithfully fulfilled the tasks for which they were designed. They changed position smoothly and noiselessly.

Unfortunately, they didn't have the time to start racing, and even less, the time to wear out. One and a half revolutions of perpetual motion. That is not very much perpetuity! It stops again. We start it up again. Same result. What could the problem be? A loupe would be utterly useless, even with very high magnification.

Then there was a further complication. I began to feel hungry, and it was time for lunch. But nothing can divert the stalwart from his task, not even his stomach; and so, patiently, I lifted the wheel from the base to inspect it again, and more or less automatically, I laid it flat on the workbench with the tip of its axis resting on the wood. Then came a surprise, a real surprise, or did my eyes deceive me? I saw, or I thought I saw a little jolt just as the axis touched down on the wooden surface of the bench. What wretched luck, I thought, the axis has come unsoldered.

I picked it up with copper-shielded pincers so as not to scratch it. It was mobile, it really moved! More than that, it turned. I gave it a quarter turn to the right. It came back all by itself! The drop of oil had produced unexpected results. It had dissolved the rust. In a flash, the truth penetrated like a bolt of lightning, and I saw that I had been even stupider than I ever thought I could be. The hub of the wheel was hollowed out to form a going barrel, and there surely must be a spring inside. At first we were too stunned to react, and then we burst out laughing. We got ready to leave for lunch, each in his own way, but first I had to pay a round of drinks.

As a matter of fact, I had promised to treat to drinks if the clock worked, but in spite of the altered circumstances,

I was very glad to offer the round, first out of fellow-feeling for my co-workers who had been just as anxious about the whole matter as I had been, and also, and above all, I wanted them to know that I hadn't been tricking them, and I didn't want to afford them too much fun at my expense.

My troubles were not entirely at an end. There was no doubt at all about the spring. We tested for this by clamping the small end of the shaft firmly in a vice. We rotated the wheel two or three times. We heard the music with which you are all familiar, that of the spring blades which pull loose when the oil is viscous. There could be no further question. The wheel promptly repeated backward the two or three turns we had made it take forward.

But this hub had been so expertly tooled, its lathed designs were so sharply etched that it was impossible to discover any trace of the cover of the going barrel. Then I got the bright idea of holding the centre of the wheel above the flame of a large alcohol lamp, and no matter how exquisitely the piece had been turned, a black circle of oil would appear which would show us where the opening or cover was. This is what transpired, and then it proved to be a relatively simple matter to remove the cover without causing any damage by simply tapping it gently with a softwood hammer. Inside we found a spring, partly rusted, 12 millimetres in diameter, and 2½ metres long.

My dear reader, my dear friends, really you must not laugh at us too scornfully. In spite of the repolishing of those pivots, which was no easy matter in view of the diameter of the wheel, in spite of all the depredations undergone by the axis, the wheel and the hub, absolutely nothing had budged. This was the source of my idea that the axis must have been fitted by force, or even brazed. There was no clue that could have led us to suspect this trickery, in view of the rigidity of this axis and of this hub. How could we have imagined that the hub was hollow?

And, there is a moral to my true story. I must tell you that on the very same evening I sent a telegram to my client, very concise, which could be translated as follows: 'Only human stupidity is perpetual, but, unfortunately, clocks never are!'

How was this clock wound up? There was no barrel arbor. The so-called arbor only extended the necessary distance to be very firmly secured to the frame, totally blocking the screws. And this is where the cunning of the constructor is best displayed. I cursed him for having kept me going longer than his clock. On the photograph (back view), a knurled button can be seen slightly below the small plate of the clockwork movement. The movement for the hands can shift upwards and downwards. It is held fixed to the frame by the knurled button. When you want to wind the clock, you unscrew the button one-quarter turn, push the movement upward far enough to disengage the wheel which corresponds to that which is fixed to the big wheel—(use your loupe—tighten the screw in order to have your hands free, and turn the big wheel backward *twenty-four or twenty-five times*, until you feel the resistance of the ratchet of the going barrel, then push the movement back down again to engage the gears. Everything is marvellously adjusted by banking studs, and the clock is ready to run again, for another 'perpetual' twenty-four hours! The ratchets and their weights provide an admirable diversionary performance, which illustrates the care and patience used to achieve this masterpiece.

As for the governor, about which I was so concerned—, it served only one useful purpose, that of exciting interest or dazzling the eyes of the investigators.

I have since learned, through an old friend, Mr Cottet, a

watch-maker with the school of Arts and Trades, that this clock had been displayed in the shop window of a watch-maker in Marseille, or thereabouts (and I say 'thereabouts' to avoid harming the reputation of anybody), and that it was advertised as being 'perpetual', and, in addition, there was a 50 centimes charge just to look at it.

I end my story by saying that since June 1910 when I repaired the clock and classified it under number 10.806, it keeps on running, it is always 'perpetual' if we remember to turn its big wheel backwards twenty-five time. It is still in impeccable condition, just a trifle oxidized or tarnished, protected by a big glass bell, and thus enclosed, it regally adorns an opulent salon.

Makers Index

Figure references are indicated in boldface.

General Index

Figure references are indicated in boldface.

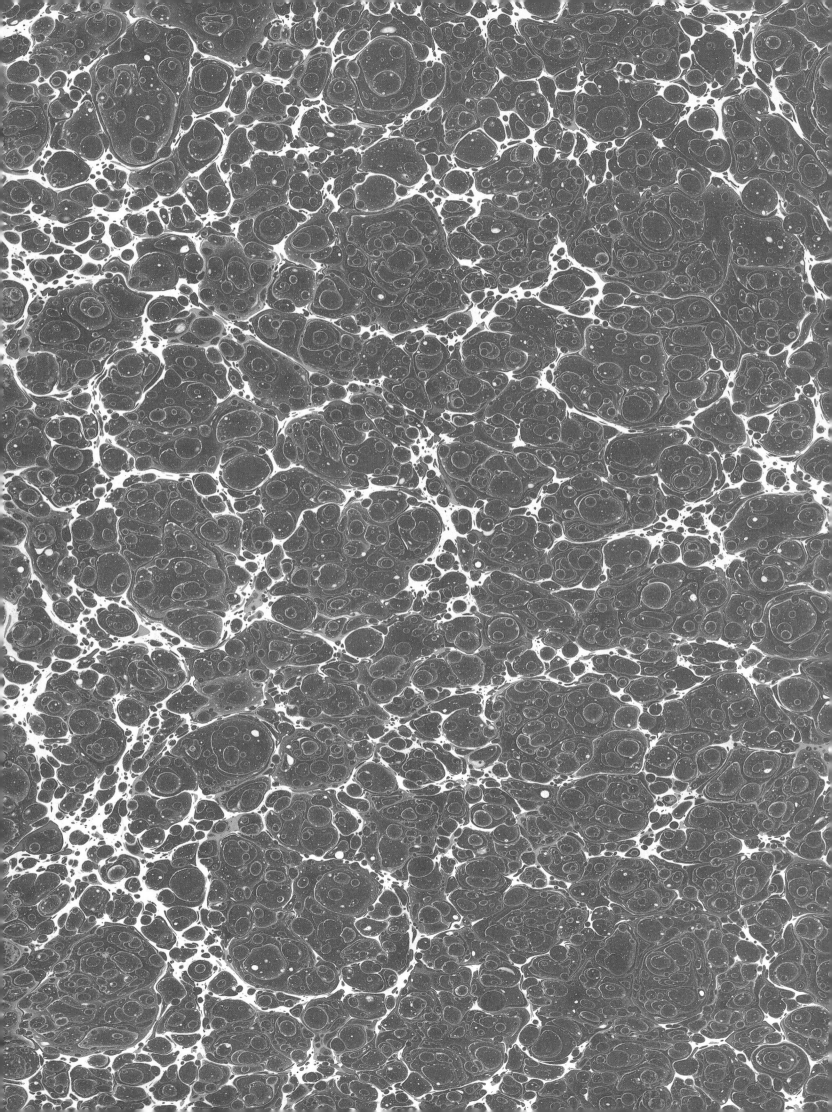